CHRISTIANITY, EVOLUTION AND THE ENVIRONMENT

FITTING IT TOGETHER

CHRISTIANITY, EVOLUTION AND THE ENVIRONMENT

FITTING IT TOGETHER

Barry J. Richardson
Foundation Professor of Biological Sciences,
University of Western Sydney

A UNSW Press book

Published by
University of New South Wales Press Ltd
University of New South Wales
UNSW Sydney NSW 2052
AUSTRALIA
www.unswpress.com.au

© Barry J. Richardson
First published 2001

This book is copyright. Apart from any fair dealing for the purpose of private study, research, criticism or review, as permitted under the Copyright Act, no part may be reproduced by any process without written permission. Inquiries should be addressed to the publisher.

National Library of Australia
Cataloguing-in-Publication entry:

Richardson, Barry J.
Christianity, evolution and the environment: fitting it together.

Bibliography.
Includes index.
ISBN 0 86840 795 X.

1. Religion and science.
2. Evolution — Religious aspects — Christianity.
3. Natural theology. I. Title.

261.55

Printer BPA

For my family,

*James, Charles and Veronica,
Christine,
Vera and Ron*

CONTENTS

PREFACE ix

CHAPTER 1 1
THINKING ABOUT THE WORLD: SCIENCE AS A METHOD
The assumptions of science 1
The nature of the scientific process 5
Scientism and pseudoscience 10
The explanatory power of science 11
The limits of science 14

CHAPTER 2 15
THINKING ABOUT GOD: THEOLOGY AS A METHOD
The assumptions of theology 16
Natural theology 19
The limits of theology 21

CHAPTER 3 23
EVOLUTION AS HISTORY
Assumptions made 23
The beginning of the universe (the big bang) 27
The evolution of stars and planets 30
The origin of life 32
The evolution of life 35
Origin of humankind 38

CHAPTER 4 41
EVOLUTION AS PROCESS
Entities and processes 41
A context (the anthropic principle) 43
Relationships 45
A history (chance, necessity and choice) 46
Death 57
Progress 60

CHAPTER 5 63
CREATION AND EVOLUTION

Freedom, purpose, chance and necessity 66
Spirit and matter 74

CHAPTER 6 78
REDEMPTION: FROM WHAT, TO WHAT?

Redemption from what? 78
Redemption by whom? 84
Redemption to what? 90

CHAPTER 7 94
OUR PLACE IN CREATION

Enabling nature 99
Intergenerational equity 100
Our role as part of nature 102
Our role as other than nature 104

CHAPTER 8 109
CREATION

A story of creation 110

GLOSSARY 115

FURTHER READING AND BIBLIOGRAPHY 121

DISCUSSION STARTERS 125

INDEX 128

PREFACE

This book is the consequence of several factors. Firstly, I needed to resolve for my own benefit the conflicts that result from confronting my professional knowledge as an evolutionary/ecological geneticist and conservationist with much that passes for theology in the modern world. I needed a sound intellectual and emotional framework that provided me with the necessary capacity to live consistently and effectively in the world. This book is a part of, and a consequence of, my reflections. It is not then a philosophy book, looking at faith and science from the outside. It is intended to show both these activities as lived from the inside. Secondly, our society has little understanding, and less empathy with what the practice of science is really like, the need for commitment to follow wherever the search leads for the truth we believe exists, and how this search affects the way one lives. Thirdly, the interaction between this and the spiritual dimensions of life seem to me to be very important, especially in our present, rampantly relativist, postmodern world. Fourthly, the assumption that the spiritual and the material are both important but unrelated parts of life made by some scientist writers also seems outlandish to me and in need of a response.

To these factors must be added a desire, resulting from my role as a member of a university, to provide information and an intellectual and moral challenge to both university students and interested lay people on the relationship and insights that can be obtained when science and religion are held in constructive tension.

While it is common to assume that there is a fundamental conflict between science and Christianity, I personally have never found this to be true, unless rather simplistic views are taken about the nature of reality. Specifically this book has not been written as a defence of evolution, the reality of evolution is taken as historical fact (though evolutionary mechanisms are still under study). Similarly Christianity, as a revealed religion, is not defended but accepted. I have taken as a given the basic structure of Christian belief as summarised in the Nicene Creed. My intention is to move on from the rather sterile and artificially contrived conflict between material and spiritual perspectives that are maintained by some, asking rather what have these fields of human experience and endeavour constructively to say to one another at the start of the 21st century. The need for this dialogue has never been greater as we face the deep, fundamentally moral, issues of this time, particularly the balance between fundamentalism and relativism in human affairs, between high technology and simplicity of life, the needs of a rapidly increasing human population and of the blasphemy of the imminent extinction of millions of species of plants and animals by our actions. This, then, is a book in the realm of what used to be called 'natural theology'.

While this book has been written from a Christian perspective, I believe Jewish and Muslim believers would, with the exception of Chapter 6, find the material considered relevant to their reflections on the relationship between science and their faith. Because it takes the reality of a personal Creator God as a given, it would be less relevant to believers from other faith backgrounds.

The book consists of four parts; the first two chapters introduce the methods of science (Chapter 1) and theology (Chapter 2), highlighting the similarities and differences in approach used. The next two chapters describe the evolutionary history of the cosmos, from the big bang to the arrival of humankind (Chapter 3) and our present understanding of the processes that underpin this history (Chapter 4). Chapters 5 and 6 consider the effects of these discoveries on our understanding of the Christian concepts of creation and redemption, respectively. The final two chapters take these insights and address two issues that face us today, the environmental crisis (Chapter 7) and the need for a 'true' myth to live by (Chapter 8).

I personally dislike books that break the thread of the 'story'. As a consequence I have not referenced or footnoted the text. Instead I have added a glossary and a list of sources for further reading, sometimes with comments as to why they are included. The further reading is given by chapter and includes, unambiguously, the sources of the few quotations in the text. The list includes works that take quite different positions from mine on most major issues considered in the book. I also need to add a further caveat: by professional training, I am a biologist and consequently the sections covering physics and geology in the book are lay summaries of concepts from these fields. Books by workers more competent in these fields than I are included in the Further Reading section.

I am grateful for the support of my own university and to the hospitality shown to me by the School of Biological Sciences at the University of East Anglia, where much of this book was written. I am also grateful to the staff and other students of St Mark's Institute of Theology, Canberra, and to the other members of the Species Survival Commission of IUCN, for their insights and friendships over many years. The book has been materially improved by the comments and critical observations made by friends and co-workers: Alan Friend, Mary Farrelly, Paul Oliver, John Cornish, and, particularly, Paul Wormell, who have undertaken the onerous task of reading and commenting on drafts of the book. I would also like to acknowledge my great debt to Christine Richardson for her loving support and insightful suggestions throughout the development of this book.

Whilst writing this preface, I have been reading T.H. Huxley, and in his autobiographical notes he says:

> ...there is no alleviation for the sufferings of mankind except veracity of thought and of action, and the resolute facing of the world as it is when the garment of make-believe by which pious hands have hidden its uglier features is stripped away.

I hope that this work shows a commitment to address the efforts of such 'pious hands' of many different faiths.

Barry Richardson

I

THINKING ABOUT THE WORLD: SCIENCE AS A METHOD

THE ASSUMPTIONS OF SCIENCE

The image of scientists is quite ambiguous in our society. On the one hand they are seen as imaginative, daring and adventurous; out there pushing back the frontiers of knowledge. On the other hand they are seen as critical, factual, remote, lacking in some human quality. Our images of science are equally ambiguous. With the fund of scientific knowledge doubling every eight years, science is a massive, unstoppable force irrevocably changing the world around us. In the process it destroys much that we hold dear and changes our perspective of ourselves, or, alternatively, it can be seen as the source of life-saving drugs and food for a starving world. Science is also a dynamic expression of human creativity and curiosity and an important cultural activity. These ambiguous images confront us whenever we think of science.

What actually is science? To a scientist like me it is simply a way of finding out about the material world in a reliable way. It functions by only accepting what can be intersubjectively verified. That is, science only accepts data that, at least in principle, can be tested by another worker. This testing would be done on the basis of the methods and experimental design provided by the original worker when reporting the results of the work. The strength of science is based on the assumption that, if two or more different individuals working at different times and places can reproducibly obtain equivalent data when carrying out the same procedure, then we can have confidence in the material facts of the observations. The interpretation of the observations however may differ. This intersubjective verification is what separates science from other fields of intellectual endeavour. Experiments that cannot be reproduced by others are not acceptable as a basis for serious consideration in the scientific literature.

Science is based on a series of assumptions that are accepted by scientists as they carry out their work. The first assumption is that the material world is not random or erratic in behaviour: it is knowable, understandable, logical and reproducible. This assumption is essential as without it the tool of intersubjective verification would not work.

The second assumption is that the universe is subject to one set of laws that can be discovered. Scientists take as an assumption that the same laws apply on a distant star, within a cell or in the remote past. All material phenomena, whenever and wherever they occur are subject to the same fundamental laws, though the form this expression takes varies according to the situation.

The third assumption is that the senses give direct knowledge of phenomena - 'something is really being observed'. If I place a solution containing pure DNA into a spectrophotometer I can obtain an absorbence reading that tells me how much DNA is in the sample. Now I may have made a mistake and my DNA is not pure. As a consequence, my estimate of DNA concentration is wrong. No scientist, however, would doubt that there really was a spectrophotometer on the bench that could give me a reading that really represented the absorbence value of my sample.

To this may be added two more concepts that have developed because they fit the observed behaviour of nature. The first of these is that there are emergent properties that are found at each level of complexity. For example, the laws of chemistry underlie all observed biological phenomena; however, the genetic processes occurring in the plants and animals found on earth and the laws that these processes follow could not be predicted simply from a knowledge of chemistry. As systems become more complex, new levels of interaction occur that obey their own, irreducible rules. This occurs without the rules of other levels of organisation being broken. The reductionist agenda, for example of describing the laws of genetics or ecology entirely in terms of the laws of chemistry, leaves out most of the laws of these sciences. Reductionism is really not an adequate description of nature. Nevertheless, all higher order phenomena must be tested by deduction to ensure that the proper reductionist agenda, which holds that nothing happens contrary to the underlying laws, is satisfied.

The second is that the laws of nature are best described mathematically. As the scientific enterprise has developed it has been found that the relationships and processes of nature can be most accurately and succinctly described in the form of mathematical equations. Even more importantly from our perspective is the observation that the best mathematical descriptions have a simple and elegant form. In fact, mathematicians will take as a working rule the practical dictum to go for elegance in developing theory as it is much more likely to produce a useful result than an unbalanced, ugly proposal. Elegance, as perceived by humans, seems to be a real attribute of the laws of nature.

Research in quantum theory has forced a further development of what scientists understand they are about. In quantum physics where single or few particles are studied, the outcomes of single experiments are not predictable. Firstly, it is not possible to measure both the position and momentum of a particle. To measure one is to make the other indeterminable. Secondly, in contexts where a particle may appear in either of two states, there is evidence that which state is present is determined when the state is measured, that is, a pre-existing state is not measured but the state itself is determined in the act of measurement. Thirdly, the act of determining states also determines the state of other particles derived from the same source atom *at a distance, and instantaneously (that is, faster than the speed of light)*. Fourthly, such particles have dual characteristics, existing, when measured by certain techniques, as waves and when examined in other ways, as particles, but particles whose position can only be described as a probability, that is, as a 'cloud' of statistical probabilities. These results raise critical questions regarding the underlying assumptions of science. Firstly, there are laws but these act as statistical likelihoods rather than absolutes (nevertheless, they can still be expressed mathematically). In sciences like chemistry where the actions studied are the average of very large numbers of events, for example a chemical reaction involving many molecules, the results are highly predictable. In both quantum physics and in biology, however, where single events as well as averages over many events are important, chance can overwhelm statistical probabilities and outcomes vary unpredictably with occasion (see Chapter 4). Finally there is a clear connection between the observer and the observed. The act of measurement affects, or in certain circumstances can determine, the outcome. The traditional view of the scientist as an observer independent of the experiment must be carefully reviewed. There is however independence, in the statistical sense, in properly designed experiments, even in quantum mechanics. While the description of science as a way of knowing that has been presented here still holds, our understanding of the nature of material reality has undergone a major paradigm shift (see below).

THE NATURE OF THE SCIENTIFIC PROCESS
The process of science consists of a set of three steps that are repeated as often as necessary. These steps can be called discovery, verification (or justification) and falsification.

DISCOVERY
Discovery is the first step and occurs through induction or intuition. Induction is the oldest part of scientific method and its identification as a recognised method goes back to the 14th century. In this process basic data is collected, and, if sufficient data is collected then patterns will be observed and, from these patterns, generalisations can be made as to the way the world works. This method depends on the orderly observation of nature. For example, a block of metal is observed to expand when heated; observation of many blocks shows the same phenomenon. When blocks of different metals, of different sizes and at different temperatures are heated the same observation is made. From this set of observations it is a step of induction to the generalisation that all metals expand when heated. Induction works by going from the specific to the general.

We now know that this method has its limits. Observations must be made in context; that is, a set of assumptions are accepted and the observations made within this framework. For example, Newton in much of his pioneering work in physics assumed that the shortest distance between two points was a straight line. Einstein in the process of his work predicted that this assumption was false, a proposal that subsequently has been upheld by observation. The most important problem with induction is that it does not lead to explanations, as it does not explain but simply observes. This method is still used in situations when we know very little. In such situations we collect and try to order our observations.

A second related approach is that of intuition. Here the scientist plays with the situation: looks at the phenomenon, pokes it, views it from different angles, compares it with other examples from experience, applies analogy to it ('This is like ...'). In a field where there is a solid body of dependable information (that is, reproducible observations) these methods are inefficient; however they are an essential part of the scientist's arsenal of techniques. Imagination is the key to the development of new insights. The nearer one works to the edge of science (rather than simply 'filling the gaps'), the more frequently we use these techniques.

As far as possible all preconceptions are identified and set aside when using intuition and we come to nature as children. The imagination is given free rein; the more adventurous the better. In these situations, when ideas come they usually arise suddenly; there is a change of perspective or a pattern is recognised by analogy. This skill cannot be learned but it can be trained and scientists have their own ways of promoting its effectiveness. Firstly, we obtain all the available facts, hand them over to our subconscious minds and wait for something to happen. If we are blessed with the right kind of mind suddenly it all comes together in a pattern, which may or may not be the right one. To help the process along, the information can be subjected to reflection and free association. The data are often discussed with other scientists, at this stage in a light-hearted way. The ideas are played with, exaggerated, turned upside down, squashed and stretched into different shapes. Wild flights of fancy are indulged. The ideas and any relevant frame of reference can be viewed from another perspective, scientists will change sides and argue for the opposite to what they have been saying. They will argue sober; they will argue drunk. They will argue late at night over the port and in the cold light of dawn. This is one of the best parts of science!

This use of imagination builds on the immediate facts and previous concepts and experience. There are, after all, an infinite number of hypotheses but these are limited to a manageable number by facts, thought frame and auxiliary assumptions. Auxiliary assumptions are those made on the basis of other sciences. For example, biologists take the laws of chemistry as a set of auxiliary assumptions, while chemists may be testing and questioning the insights underlying some of these laws.

One of the challenges at this stage results from the need to use previous concepts to limit the almost infinite range of possible hypotheses. This must be done, however, while knowing that at least one of these preconceptions (probably one we did not even realise we were making) is wrong. Science has sometimes been called 'hunt the incorrect and unrecognised assumption'!

As the introduction of intuition extended our capacity to develop new concepts based on the use of induction, so the development of new statistical methods, in concert with the availability of powerful computers, has further extended our range of approaches. For example, the use of multifactorial analysis provides us with new ways to identify patterns in very complex data sets. We can also analyse several models at the same time using maximum likelihood analysis to measure the relative probabilities of the alternate models.

VERIFICATION

In the second phase, verification, the new idea or insight is tested against all that is known. To do this the light-hearted approach used in the first stage is put aside and a disciplined mental approach is used. We mentally draw out the idea and test it against known facts, looking for inconsistencies. In this stage deduction is used: that is, we now argue from the general to the specific. The world is assumed to follow the rules of logic and the logical outcomes of the insight are deduced, predictions made and tested against the body of available knowledge and theory. Does it explain/predict the known facts? Does it fit with all other available facts and auxiliary assumptions/theories? Does it follow logically from given assumptions? At this stage most insights fail: the present body of knowledge is such as to show that the original insight is wrong, that is, inconsistent with available observations. We must therefore try to find another insight by returning to the beginning or perhaps by building on the nature of the failure of the original insight.

It is this phase that commonly forms the introduction to scientific research reports. They are prepared as though the insight was originally deduced from theory and this misleads people about the real process scientists follow.

FALSIFICATION

Testing or falsification is the final step in the process. The aim is not to confirm the idea by collecting instances of agreement, which is a form of argument followed in many other fields of human endeavour. Rather, the aim is to seek to falsify the idea by finding cases that do not fit. This is the usual experimental phase of research. The process starts by making a prediction from the idea. The experiment is then carried out to see if the prediction holds. The more unexpected or counter-intuitive the prediction, the more confidence we can have in the insight when the prediction is upheld. This is an example of the use of the old adage 'the exception proves the rule'. In modern English this means 'the exceptional or extreme situation tests the idea', not as it is usually interpreted. Thus, the more extreme the test or prediction that is upheld, the more confidence we can have in the idea.

If the idea is shown to be wrong, that is, falsified, then the original idea must be discarded or re-examined. We can be sure of this though: if an idea is soundly tested and shown to be wrong then it is wrong and we have at least learnt something. The opposite however is not true: if our idea is upheld by the test, we have not proved it to be true, for other ideas may also explain the data. Thus science advances not by finding what is true but by discarding what is false. The body of scientific theory that our modern technological society is built upon, consists then of a more-or-less consistent set of ideas that have withstood serious testing. The central ideas, those in which scientists have the most confidence, are those that have faced the most testing and not been disproved.

This commitment to falsification is one of the major factors that underlies the ambiguous image of the scientist in the community. A scientist should never give up the search for ways to disprove a favourite insight. A certain dispassion is needed so as to keep this commitment, for in the end all our ideas will be proven wrong, or at least incomplete. At the same time passion is needed to push the mind and the world in the search for ways to find new patterns and test the deepest insights available. It is this ambiguity that outsiders see when they look at scientists. Scientists need to be childlike or ruthless, passionate or dispassionate at different times in the science process. All beginners need to be warned however that the search for knowledge is very exciting and highly addictive!

Another problem that faces scientists is the likelihood that there will be many potential hypotheses that could explain the available information. To decide which to test first, scientists use a rule of thumb that has been found effective. This is based on the application of Occam's razor, that is *entities shall not be postulated without necessity*. Thus the first hypothesis to be tested is the one that makes the fewest assumptions. More complex ideas are only developed and tested when the simpler ones have been falsified. That is, the most parsimonious idea is preferable because it is the most explanatory and the least *ad hoc*. More complex hypotheses may result, not only from preconceptions in science but from religious or political imperatives.

An interesting sidelight that arises from following this method is the discovery that not one in a thousand of the good ideas developed during the discovery phase survive the testing process. This experience leads many scientists to view much of the theory developed in other areas of intellectual endeavour with a critical eye. Without the almost infinite range of opportunities to test ideas with an effective, intersubjectively verifiable, critical apparatus, the basis for sound, generally acceptable theory becomes very difficult. Postmodern criticism of the morass of ideas found in many discipline areas that result from the status of individual practitioners, cultural fashions and the need to say something different from everybody else, is directed at the same problem, though from a different perspective. Because of the use of intersubjective verification many of the criticisms raised by postmodernists have little to offer of relevance to a consideration of scientific methods, results or insights, though they may have more to say about the way scientists develop or present their ideas.

SCIENTISM AND PSEUDOSCIENCE

Scientists, then, can be recognised by their commitment to falsification. The language of science however has been used by many for their own ends because of the respect, or at least awe, in which science is held by the community.

The first to fall prey to this temptation were we scientists. It proved too much of a temptation for many of us to admit to limitations in the products of science or in the contexts in which science had something relevant to say. We found it almost impossible to say in response to a public request 'I don't know' or 'I cannot help'. This 'hubris', as the Greeks called it, has led in part to the tarnished image of science in the eyes of the wider community. Science as a branch of human knowledge can only address a limited range of questions and, even within this range, limitations to our present knowledge restrict our capacity to properly advise the community that supports our work.

Scientism is related to science but turns it into a form of religion. Scientism assumes that science is the only true form of knowledge. Scientism claims that objectivity, neutrality and rationality are essential, thereby discounting the role of subjectivity, emotions and intuition in human affairs. Even the process of science that I have described above would not be acceptable to such a religion! It is more often preached by those who claim to know about science than by practising scientists. It is rare indeed to find the home of a scientist not replete with books on history, the arts and other cultural matters.

Pseudoscience uses the language of science and the concepts of science to build theories. These are usually supported by examples, often mounds of examples, but there is no commitment to falsification. In fact the ideas are usually in a form that does not allow the possibility of falsification. In science a proper hypothesis must be, at least potentially, falsifiable. Such pseudoscience comes from more or less reputable intellectual equivalents of snake oil salesmen or people who are muddled in their understanding of what science is. In a second form, pseudoscience is often a helpful way of popularising scientific concepts, and many of the programs passing for science on television and in popular books are examples of pseudoscience. While informative as to the insights gained through science they are quite misleading as to the nature of science. A third way in which ideas developed in science are used is as analogies in other fields. Ideas like chaos theory, the uncertainty principle and relativity have provided fruitful starting points for the imaginative development of other fields of human endeavour. Unfortunately the fact they are used as analogies is often forgotten and the authority of science is claimed for these new uses. For example, mathematical concepts are often gravely (and sometimes laughingly) misunderstood.

THE EXPLANATORY POWER OF SCIENCE

Science is the result of the work of many individuals over a period of several hundred years. A comparison of the work of early scientists with that of modern scientists shows that our methods have been extended and our insights into the procedures we use have deepened. These insights have been further extended by sociological studies of working communities of scientists. Such studies have shown that scientists are mere mortals and that their behaviour in communities is much the same as that of other human groups. By a reasoning that surprises the scientific community (for we know what perfectly ordinary humans we are), this has been taken to cast doubt on the results of science. For example, it is quite clear that there are fads in science. There are also areas where the development of ideas or technology make rapid progress possible. This year meiotic drive is all the rage in genetics, next year the genetics of peripheral populations holds the stage. Scientists recognise this phenomenon and even have a name for it, it is called 'bandwagon science'. These are the subjects getting the grant money this year, hop on the gravy train. It is all very human. However, the work that is carried out by those working in popular areas is usually at one of the frontiers of science and is done according to the canon of science. The fact that effort is concentrated in one area rather than another really has nothing to say about the validity of the results obtained.

A deeper question comes from studies of science over time. It can easily be seen that science is not a unity but a series of interlocking ideas, and a critical test of the validity of a particular idea or set of ideas is how coherent a picture it makes when combined with other scientific insights. Underlying these sets of interlocking ideas is one or several sets of fundamental concepts, or paradigms. All reputable science must fit the dominant paradigm. Often this paradigm is implicit rather than explicit but anyone who bucks the system is 'unsound' as Sir Humphrey Appleby of *Yes Minister* fame would say. Over time, however, there is a gradually increasing number of cases of misfit between theory and facts or between theory in one field and theory in another. A store of anomalies builds up to the point where it cannot be ignored. As I described earlier, scientists love and hate anomalies because they test and falsify our ideas. As Winsor's revision of the old nursery rhyme says:

> This is the theory that Jack built.
> This is the flaw that lay in the theory that Jack built.
> This is the erudite verbal haze that covered the flaw that lay in the theory that Jack built.

Ultimately the load becomes too great for the dominant paradigm and someone comes up with a new way of understanding the data. A new paradigm comes into existence. It is then tested and refined and we work to extend its range of validity. Eventually the cases of misfit begin to accumulate again until we face another intellectual revolution. Some would hold that this places a serious question mark over science as a way of knowing, for an outdated dominant paradigm often holds on for longer than it should in a perfect world. It must be remembered however that the old paradigm worked quite well at predicting the way the world functioned over a wide range of conditions, though not all. Newton's physics is still perfectly adequate for determining the trajectory of rockets, and the more general theory that superseded it, for example the special relativity theory of Einstein, includes Newtonian physics as a special case. The turmoil at the time of scientific revolutions reflects the very human nature of scientists and power structures, not the adequacy of the data or the incomplete nature of the theory.

What then constitutes satisfactory theory? Clearly it is theory that works. It is theory that predicts the correct outcome over a range of situations. It is theory that makes a coherent picture when combined with data and theory developed elsewhere in science. A better theory would be one that made fewer assumptions or worked over a wider range of conditions or was more closely integrated with theory in related fields.

This description is quite different from the assumption made earlier where I said that science aims to find and describe the fundamental laws of nature and show how they bring about the complexity and diversity apparent in the material universe. Scientists would say that a satisfactory theory is an approximation to those laws: better theory is even closer to those laws. Further work will uncover anomalous cases that will ultimately allow us to build theories that are even closer to the real laws. Whether our aspiration to find the real laws will ever be met or we will only continue to develop closer and closer approximations remains to be seen.

THE LIMITS OF SCIENCE

What then do we make of science theory as a description of the world? Here we have to return to the basis of the scientific method. Satisfactory scientific theory tells us what the world is *not*. The more satisfactory the theory the wider the range of options about the nature of the world that we know are not true.

Perhaps such a picture of science is not as glamorous as pseudoscience or scientism would like us to hold, but at least we know what we know!

Science, then, provides us with a great deal of ordered descriptive data about the nature of the material world and it provides a great deal of information about the way the world does not work. It also provides a set of predictive rules as to the outcome of a range of actions under a range of conditions that, in some sense, approximate the laws underlying the material world. The range of science is limited to intersubjectively verifiable material facts and their interpretation. Beyond this it has little to say in a positive way. It can however test theories raised in other areas of human endeavour in so far as they lead to predictions in the material world. But more of this later.

2

THINKING ABOUT GOD: THEOLOGY AS A METHOD

The image of theology in Western society has suffered a fate even more extreme than that of science. From a standing at the centre of the intellectual and cultural life of the community in medieval times, it has simply become irrelevant to the directions taken by modern, Western society. Nonetheless, surveys of Western communities show that most individuals still believe, though in what they find difficult to say in any clear fashion. In this chapter we will briefly review the basics of a theological method and consider the possibility of natural theology, that is, using our knowledge of the natural world in theological reflection. This chapter deals with methodological issues, not the content of Christian doctrine as such.

The steadily increasing capacity of humanity, through science, to control the forces of nature has not been matched in society by an increased sense of responsibility or any sign of an increased capacity to act in a morally responsible fashion. It is in this context that we see individuals, in response to a recognition of the reality of the moral and spiritual dimensions to life, react to modern, materialist society by undertaking an earnest and heartfelt search for spiritual truths by which to live. With no trust in the capacity of the traditional Christian churches to act as guides or mentors in this search, or worse yet, with a real sense of disillusionment that follows when they do approach such churches for adult guidance on the realities that confront them as mature men and women able to build adult, responsible lives in the world, they reinvent theology for themselves. The resulting primitive theologies often have been built in ignorance of the insights of two thousand years of theological reflection, due to the failure of the church to meaningfully present its insights or, alternatively, to renew its insights in the light of a world whose understanding has been transformed by the work of science. This failure stands as an indictment of the church. The old adage that the only alternative to good theology is bad theology is evident in many different ways in our time.

THE ASSUMPTIONS OF THEOLOGY

What then is theology? It is literally 'the study of divine things' and it aims to investigate Christian revelation and what Christians believe by the use of reason in the light of faith. It aims to clarify and deepen our understanding of our faith. Thus, it is not the experienced faith of Christians but an examination of this experience, perhaps, more truly, a dialogue with it. Theological reflection takes place within the Christian community and is a consideration of revelation, that is, of God's dealings with the world. It is not theology to study Christianity from the outside, that is a branch of the philosophy of religion. Theology is the result of reflection on a lived/experienced reality that reveals itself.

Like science, theology is based on a series of assumptions. There is, however, nothing like the consistency of assumptions or methodology in theology that is seen in the natural sciences. Like science, however, theology wishes to make explicit that which it studies, and both have to grapple with mystery.

The first assumption is that reason, properly applied, may be validly used to gain insight into the nature of divine things, in so far as they have been revealed. Similarly in science, reason is applied to the material world to reveal its underlying nature.

The second assumption is that there is material or data to which reason can be applied. In what form then do we find God's revelation, or action in the world, that we may reflect upon it? Clearly, the experiences of God that we draw upon have many dimensions. Firstly, there is our personal history and experience. Our lived faith always provides the context that underlies our theology. Any valid theology must be consistent with our personal experiences be they religious, moral or, more generally, of the world. A personal theology that either ignores or contradicts our life experience can hardly last.

Secondly, it draws upon the personal experiences of others, and especially the combined and integrated personal experiences of others. That is, it must be consistent with, and draw from, the shared life experience of the Christian community. As we have a personal set of experiences, so they can be examined by reason and so we can place them and test them in the wider experience of the many Christians in the past or present, and can take account of their insights into this shared experience. This pool of experience and insight is wider than we can experience in many lifetimes and any valid theology must be able to account for this fund of experienced faith and life.

Thirdly, this pool of experience must be interpreted in the light of the experience of the first generation of Christians who shared the earthly life, death and resurrection of Jesus. For Christians, the person, teaching, actions and continuing life of Jesus constitute the central basis of faith. It is the experience of this reality, linked through the common thread of the life of the Christian community and shared through the ages, that gives a peculiarly Christian spiritual and moral meaning to personal experience. It is these primary events, this revelation of God in a human life, that has shaped the faith community of Christians and gives it its basis.

Fourthly, this primary insight of the first Christians is available to each generation in the text of scripture. This means that each generation, including the present one, has directly available to it the descriptions and insights of the experiences of the first and second generations of Christians. These different sources of information allow the present generation, by comparing them, to test the validity and value of the Christian faith and interpret it for itself.

The work of theology therefore involves reasoned analysis of God's actions as revealed to each generation in the living experience of individuals. This experience consists of their personal experiences, in the integrated experience of the community of faith and in the interpreted experience of the earliest Christians. Any valid theology must provide a consistent interpretation of the whole of this body of material.

The task of theology, then, consists of the identification and consideration of any and all aspects of reality (things, events, experiences; past or present) that may be considered the results of God's revealing actions wherever they may be found. These must be collected in an orderly fashion similar to the discovery phase of science and, using the same tools of induction and intuition, insights must be developed from this material as to the nature of the underlying reality reflected in the material. As with science, imagination plays a critical part and as far as possible cultural assumptions are put aside. As with science, with its base of intersubjectively verified data, so with theology: though here the data is the orderly documented experience of the individual and the community.

As in science, a series of auxiliary assumptions are available and used in theological reflection. These consist of the set of derivative ideas that are generally accepted by Christians and used to interpret personal revelation. These would, for example, include belief in the Trinity and the concept that the books now found in the New Testament are fundamental to the faith. Again, as in science, auxiliary assumptions need to be made explicit, and, in other parts of theology these auxiliary assumptions may be subjected to critical analysis. The concept of the necessary coherence of the whole body of theology, if it is to be considered valid, and an aggressive attempt to address any inconsistencies, are also parts of theological method.

The second stage is again like that of science where a new idea or insight is tested against all that is known. Thus the process of verification consists of insights that are tested against all available relevant information. Deductive logic is applied to the ideas, and the conclusions reached are also tested against available information. Here, for example, the generality of an insight gained by Christians living in one culture can be tested against the experienced faith of Christians living in other cultures (either in another place or time), or the experienced faith of men can be tested against the experienced faith of women.

There is no equivalent to the third stage of science, falsification, unless some form of revelation or previously unknown attribute of revelation can be predicted and then data generated to test the prediction. In theology, work cannot go beyond the second stage and this is usually done by providing many examples from as wide a range of situations and authorities as possible to support the proposal made.

It follows from the earlier argument that theology is concerned with the examination of the revealing acts of God. One set of such acts are those that led to the present material world, to the laws that it follows and the nature of the structures found in it. This is the subject of present-day natural theology.

NATURAL THEOLOGY

Natural theology is that part of our knowledge of God which can be obtained by the application of reason without resorting to information or knowledge revealed through religious experience. In the Middle Ages natural theology included, for example, attempts to produce logical proofs of the existence of God. This approach has been challenged on many fronts and has been largely set aside by theologians. Interestingly, some scientists have recently taken up the cudgels for natural theology but with a far more restricted agenda. By their very nature scientists seek a consistent and coherent world view. In such a situation, unless they are willing to be intellectually dishonest, scientists, who happen also to be Christians, have to find an integrated and consistent vision of their faith and their science. We cannot take those pieces of reality that suit us and ignore the rest. We cannot be satisfied with pretending things are the way we would like them to be; we must address awkward and inconsistent facts, just as much in the relationships between faith and science as we do within science where inconsistency is the goad for renewed effort.

If the world is the creation of God then it is part of the revelation of God and a source of information on the ways of God and therefore of God's nature and intent. Thus the new natural theology does not aim to explain everything on the basis of reason but to identify and describe the realities of the creative and sustaining work of God as presented in the nature and process of the material world. These may tell us something of the nature of God but, like much scientific theory, they are more frequently going to tell us more about what God is not rather than what God is. They will help us discard concepts that are not consistent with material reality.

What then does science have to say about revelation? Not a great deal really, as most dimensions of the Christian revelation are not available to inquiry using the methods of science. There are some areas of interest however. In so far as the world is the result of the creative acts of God, it reflects the nature, intent, purpose and methods, that is the mind, of God. Thus as the insights of science develop, as we understand better the history of the universe and the processes that underlie its functioning, they extend our knowledge of the acts and methods of God and, if used carefully, can open or close options as to the nature of God's action in the world and therefore, indirectly, of the mind of God relative to the world.

There is a further reason for a renewed interest in natural theology in our society. The material culture and the intellectual culture in which we live is, to a large extent, the product of the scientific revolution. Our world view bears little resemblance to that of past generations. We still need however to make sense of our lives. We still need to make sense of the issues of life and death, of meaning and purpose. To do so means we must make sense of the world described to us by science. We must make sense of its origins, its history and its processes. Our generation also faces, as a consequence of the power of science, a series of moral dilemmas, for example, in our relationship with and behaviour towards nature. We urgently need to find a moral framework, as much as a technical one, in which to make environmental decisions.

THE LIMITS OF THEOLOGY

Theology is limited in three ways: firstly, by the nature of the data available for reflection. Much of this is highly personal and subject to the frailties and misunderstandings inherent in human existence.

Secondly, it is limited in its methods of analysis. Much of the criticism raised by postmodernists can be directed with good purpose against theology. It is culturally determined in many ways and many of its assumptions are implicit and beyond testing. Much of what is normally considered theological insight is easily seen to be largely an expression of the culture of the particular community. Reason applied in circumstances where there are large numbers of unrecognised assumptions is a perilous tool at best.

Thirdly, as is found in all the interpersonal relationships humans enter into, the subject of such an analysis is more knowable to the heart, that is, in and through the relationship, than to the mind. Given this, and also the very nature of God, there must always be a great deal that is beyond human understanding in the mind of God.

Even in the light of these restrictions, however, there is still room for constructive work. For example, there is a need to deal with ideas about God that are wrong (that is, inconsistent with the facts) and by continued iterations of reflection, concept development and data analysis we, like the scientist but with a great deal more difficulty, can make closer and closer approximations in our understanding of the nature of the mind of God.

3

EVOLUTION AS HISTORY

The word 'evolution' has so shaped our culture and thinking in the past 150 years that we must be very careful how we use it, for the word has far outgrown its original frame of reference. While in genetics it has come to mean simply the duplication of genes and changes in gene frequencies, for most people it is related to the history of the earth and for many it is used as an explanation, or excuse, for various kinds of biological and cultural behaviour. In this chapter we will look at a summary of the results of scientific research into the evolutionary process when seen as a series of historical events that have occurred and that have led to the present state of the universe. In the following chapter we will examine the mechanisms and processes involved.

ASSUMPTIONS MADE

In describing the material world in evolutionary terms we make a series of assumptions, and it is essential that we are satisfied with these before proceeding further. Firstly, we must accept the assumptions made in Chapter 1. The first assumption is that the material world is not capricious: it is knowable, understandable, logical and reproducible. The second assumption is that the universe is subject to one set of laws that can be discovered. The third assumption is that the senses give direct knowledge of phenomena, 'Something is really being observed'.

In the present context these assumptions lead to what is known as 'uniformitarianism'. That is, that the forces acting in the past that shaped the present material world are the same forces we see working now. We would expect to see in the geological history of the planet, therefore, the same geological forces acting in the past as act now. Thus erosion, volcanic eruptions, and so on, all of which can be observed today, would be found in the past. As we step back further in time, however, while we would still expect to see the same laws in action, conditions differed greatly from those we find now and we would expect to see their expression as quite strange, for example, the origin of the chemical elements. The technique scientists use in such arcane regions of inquiry is to develop a model of what they think happened, to fill out that idea to see what effects it would leave in the present, and then look to see if these effects can be found. Clearly, the predictions have to be other than the facts that lead directly or indirectly to the idea in the first place.

It can be seen from the first assumption that the aim of scientists would be to describe and explain the present structure of the universe without calling on any capricious acts in this process, that is, law-breaking events, miraculous or not. The universe should be explained in its own terms as far as possible.

This assumption limits the number of general explanations available. There are three major alternative explanations of the present structure of the universe in Christian circles. Firstly evolution, secondly, an act of God giving the universe its present form, and thirdly an act of God followed by a worldwide flood.

The last proposal was developed by treating the text of the early part of the book of Genesis in the Bible as a literal description of historic events.

The first explanation, evolution, is a central theme of the remainder of this book. At this stage it is simply worth noting that evolution as a description of a historic process is compatible with both theistic and atheistic faith systems.

The second explanation assumes that the world was formed by God relatively recently with an apparent history included, though this history never happened. For example, light takes many billions of years to travel from a far galaxy to earth, but at the moment of creation God created suitable photons of light along the path they would have travelled, if there had been a long history. There were no dinosaurs, prehumans or other fossil species; the fossils were placed in the rocks, and the rocks placed in layers as if there had been a long evolutionary history. This approach cannot be overturned by science, as all evidence, whether it is radiocarbon dating or the fossil record or the time it takes light to travel from the stars, is all consistent with a long evolution of the universe. The only objection to such a scenario is theological. Does such an explanation match our experience of God? As Einstein said, 'God is subtle but not malicious'. In Christian terms, is it likely that, having gone to the trouble of creating independent creatures, God would destroy their independence by doctoring the evidence? I would find it quite difficult to trust such a God.

The third explanation suffers from all of the problems of the second explanation with respect to the apparent length of history. The fossil beds and strata of the geological record, however, are explained by a worldwide flood at the time of Noah. The problem of the light from distant stars and other time-dependent phenomena remains. In such cases it has been argued that the universal constants, for example the speed of light or the rate of radioactive decay, have changed over time. There is absolutely no valid scientific evidence for such claims and overwhelming support for the contrary view. The second claim, that of a worldwide flood, can also be examined using standard methods. This has been done many times and in many ways, always with the same outcome. The geological and palaeontological records cannot be explained using such a system. Trivial examples of the kinds of problem that such an explanation faces are:

The geological record of many areas, for example southern Australia, includes thousands of metres of strata of sedimentary rock, the kind that is usually ascribed to the flood. In the middle of these strata are layers of aeolian sands, that is, 'fossil' sand dunes of the kind formed by wind on land in a dry environment.

In the Midwest of the United States, on the top of many thousands of metres of sedimentary (flood) rock there is a fossil coral reef (like the Great Barrier Reef). This reef is several hundred kilometres long and more than a hundred metres thick. The base of the reef grows out of the underlying sedimentary rock as it would if the coral had grown in this position. The reef would take at least thousands of years to grow to its present size and is found in the middle of what is now a continent.

There are, conservatively, 8 million species of plants and animals found across the face of the earth; many of these are found only in a restricted part of the world, often in a single small valley or on one mid-ocean island. All need particular habitats and food sources if they are to survive. Most terrestrial species cannot survive, at any stage, long term immersion in water. Many species of animal are restricted to a single food source, either a particular plant species found only in a limited geographical area, or another animal species. Many of these species, for example wasps, also live only a few days in the adult form and must find and paralyse a host in which to lay their eggs. The parasitised host animal of course never breeds.

Humans and all other complex animals are hosts of a wide range of species-specific parasites and other disease organisms.

Many of these disease species cannot live for any time outside of their host and survive by direct spread from individual to individual. An example of the many hundreds of such species is measles. After several weeks the host develops immunity to the disease. which is destroyed and the individual recovers. In the meantime however the disease organism has spread to another host that has not previously been infected with this disease. All of the humans and animals in the ark must have each been carrying hundreds of such diseases and insufficient time was available for new hosts to be born and grow up to carry on the infections.

Dendrochronology, that is, determining the age of a piece of wood from the pattern of differences in thicknesses of the annual tree rings it contains, has now advanced to the point where the sequences of thicknesses laid down year-by-year have been identified for a period covering about 9000 years. Independent sequences are available for North America and for Europe. Pieces of wood so dated can then be used to confirm the adequacy of some of the other dating methods, for example radiocarbon dating.

It is hard to disagree with the statement 'I do not believe in evolution, nor in gravity. They are facts of life. I believe in God'.

The evolutionary history of our world can be summarised as a series of five steps:
- the beginning of the universe (the big bang)
- the evolution of stars and planets
- the origin of life
- the evolution of life
- the origin of humankind.

THE BEGINNING OF THE UNIVERSE (THE BIG BANG)

Some time, roughly between twelve and fifteen billion years ago, the universe came into existence. This event is usually called the big bang because all the energy and laws that constitute the universe winked into existence as an almost infinitely small, infinitely dense, point. This universe then began to expand at the speed of light. Evidence of such an event is found in many forms. The details of the theory, however, are still incomplete. Like all science, it is 'work in progress'.

The universe is very large but one is entitled to ask if it has any edges, no matter how far away. What is proposed however is that the universe is a closed system. The furthest point in any direction is where you started. An analogy could be seen in the older forms of computer games where the plane or character could move off the screen in any direction, but as they did so they moved onto the screen on the opposite side. It is as if the screen is continuous in every direction and movement in a single direction leads one back ultimately to the same place. The universe is seen as similar but in more dimensions than the two that make up a computer screen. While the distance that can be moved in any direction is infinite, at the same time the screen is clearly finite in size, as the universe is thought to be. In each case they are finite but unbounded.

One line of evidence of the reality of the big bang follows from the second law of thermodynamics, which says that entropy cannot decrease in an isolated system. Thus the universe, considered as such a system, is steadily becoming more disordered. That is, the total stock of order in the universe is steadily being used up. It follows that the universe must have had a beginning, since, if it had been here forever, it would have come to equilibrium.

A second line of evidence arises from the observation that galaxies are moving away from one another, that is, the distance from here to a far galaxy is increasing. Even more surprising is the observation that the speed at which galaxies are separating is greater the further apart they are. The universe is expanding! It is therefore possible to work backwards and calculate an approximate date for the time when all the matter in the universe occupied the same point. This date is of the same order as that estimated for the age of the universe by other methods: between twelve and fifteen billion years.

An important concept is that the big bang was not a single moment when matter started to expand to fill the empty regions of space around it, for the power of the gravitational field of the point was so great that space itself was drawn into the point. It was as if the screen in our television game was contracted to a smaller and smaller size until it became an infinitely small point. To characters in the computer game there is nowhere outside the point. In the reverse direction then, after the big bang the 'screen' started to expand in size and all the points on the screen steadily grew further apart. From any point on the screen all other points seemed to be moving away. In all cases however, no matter how large or small the screen, after the first moment it would be possible to travel an infinitely long way. In the real universe the rate of expansion is the speed of light, and, as matter cannot travel faster than the speed of light, it is not possible to travel across the universe as it is expanding too quickly.

A third line of support comes from the analysis of the conditions likely to have occurred during the first seconds and minutes of the life of the universe. At temperatures of many billions of degrees, atoms as we know them could not exist as their electrons would all be knocked off. There were, however, nuclear reactions between the hydrogen nuclei. These would fuse at such energies to produce helium. It is possible to calculate the expected ratio of hydrogen to helium that would result from these collisions and the time and conditions of the early expansion phase of the universe. The expected ratio is approximately three to one, as is found in the actual universe.

A fourth line of evidence follows from the work of Einstein. Whilst working on the general theory of relativity he attempted to solve the equations he had developed to obtain a mathematical description of the space–time physics of the whole universe. At that time it was assumed that the universe was stable, as the expansion of the universe remained to be discovered. However, his solutions showed that the universe should be expanding and he was forced to include a 'fudge factor' to remove this effect. Five years later the equations were solved by another worker without the fudge factor, thereby demonstrating the expansion. Einstein always believed this to be one of the most serious blunders he ever made. It is always a challenge in science to step outside the set of preconceptions that one has, to truly explore the nature of the universe.

Time becomes involved in the beginning of the universe because time and space and mass are all intimately related. If space is contracted to a point, time also collapses. Or to see it in the opposite direction, at the moment space starts to expand, time also comes into existence. There is no time before the big bang. Creation does not start at a particular time. Speaking theologically, God did not think about initiating the universe before it happened because there was no before. The existence of God is timeless, that is, outside time; the entire history of the universe is comprehended by God as a single unitary experience.

THE EVOLUTION OF STARS AND PLANETS

A few hours after the big bang the reduction in temperature due to the expansion of the universe reached a stage where the production of helium by nuclear fusions stopped. The conditions of the universe had now reached the stage where atoms could remain in existence, as their electrons would not be continually knocked off. Gravity now began to take a hand in the process. Small chance differences in the density of matter led to focal points of gravitational attraction. Though the universe overall was expanding, matter in local areas began to be drawn towards common centres at rates faster than the expansion rate. These steadily increasing concentrations of matter became protogalaxies. Within them, on a smaller scale still, the unevenness in the distribution of matter led to further focal points, and these ultimately became stars.

As these clouds of hydrogen contracted, the temperature rose until it was high enough for nuclear fusions to occur, as was found in the first few minutes of the life of the universe and as happens within a nuclear explosion or in an experimental particle accelerator where these reactions have been studied. In these first generation stars, hydrogen was burnt to helium and then, by further fusion reactions, to larger atoms containing many protons and neutrons. In explosions late in the life of these stars, the heaviest elements were formed and matter was flung back into the wider galaxy, only, in combination with more hydrogen, to contract and form new, second-generation stars. The continuation of this cycle gradually changed the proportion of hydrogen to heavier atoms. The sun, born five billion years or so ago, has 2 per cent heavier elements. Thus the oxygen, carbon, iron and other elements of which our bodies are made were formed in stars long ago. We are literally part of the long and continuing history of the universe.

Because the cloud of gas that condenses into a star is spinning, a balance is reached where some of the material does not fall into the star but ends up orbiting the star. Again gravity has its way and planets form by mutual attraction within this cloud of material. Planets, however, are too small for temperatures to reach the levels needed for nuclear fusion to start, and, though hot, they are not hot enough. Studies of nearby stars have detected the presence of planets circling several of them.

The results of detailed studies of the life histories of stars, nuclear reactions within stars and in laboratories, plus the examination of galaxies many billions of light years away (that is, as they were billions of years ago) support the current scientific model of the differentiation and production of the universe as we see it. Our understanding of this history and the processes underlying it, however, is still far from complete.

An interesting question rises from this process, for the final state of the universe seems on first impression to be more structured than the first state. The increased state of organisation, with galaxies, stars and planets, means that the second law of thermodynamics seems to have been broken. Now the second law only holds for isolated or whole systems. Thus it is possible to increase complexity in one part of a system by increasing the entropy of other parts of the system. In the solar system, for example, increased chemical complexity on earth, in the form of plants and animals, is primarily 'paid for' by increased entropy in the sun (through, among other processes, the transfer of energy in the form of sunlight). This does not seem to have been the process used as the galaxies were developing. The driving force of these events was gravity. For matter at scales too small for gravity to be important, high entropy means high disorder. Thus if two chambers, one containing nitrogen and another containing oxygen, are joined the gases will diffuse through the combined space. The structured state of the two gases in separate areas will be lost, and entropy will increase. The opposite is true for situations where gravity is significant because disorder of material objects is only one aspect of entropy. Another aspect is related to the distribution of energy in a system. Conversion of gravitational potential energy into heat also causes an increase in entropy. At large scales the high entropy state is one where the matter is drawn together into lumps, that is, differentiated into galaxies and stars.

THE ORIGIN OF LIFE

The origin of life is one of the most interesting areas of current research into the whole historical process, as we know so little about it. Part of the problem results from the difficulty in defining exactly what we mean by 'living'. All living creatures currently found on earth are extremely complex and have a great many attributes in common. This is due to their derivation from a common ancestral lineage that lasted for perhaps half a billion years. It is easy to define life in terms of the attributes of the most recent of these common ancestors. However, the first organisms in this long lineage must have been much simpler than anything living today. We would see these first organisms as only half alive, much like a virus which can divide and evolve in the right environment, but can also be crystallised out of solution in the same way as a simple chemical compound. The present highly complex cellular system of a store of genetic information in the form of DNA, which is converted into a set of instructions in the chemical form of RNA, is used to produce a wide range of proteins with particular functional properties, clearly the end product of a long and involved process of evolution. The first organisms must have been much simpler. The key step needed to get the system of living organisms established was the production of a relatively small self-replicating molecule with the capacity of facilitating the production of copies of itself.

In trying to imagine the processes that occurred, we need to take account of the conditions that were present at the time. It is difficult to grasp the enormous changes that have taken place since those times. For example, the atmosphere of the newly stabilised earth contained effectively no free oxygen but consisted of gases similar to those vented from volcanoes; lots of carbon dioxide, and other gases. Studies have shown however that such mixtures, when treated with high temperatures, electrical discharges (lightning), or ultraviolet light (freely available before the earth's ozone layer formed), will form a thin soup of organic compounds, including amino acids, nucleic acids and other compounds found in living organisms. Astronomical studies have also shown that many of these compounds are present in outer space, in the dust clouds that occur throughout our galaxy and elsewhere. Thus as the earth condensed, many of the compounds that today we associate only with living organisms would have been present already, with the stock added to over time by the action of lightning, UV, and so on. If such a soup was to occur today it would immediately be broken down by bacterial action; however, at the time there were no organisms present to use these resources.

A range of possibilities has also been proposed for further concentration of these simple compounds; for example clay can absorb simple organic compounds by binding them to the charged surface of the clay. The presence of metal ions in this environment can then cause polymerisation of the monomers, that is, they can link up to form long chain-like molecules, as amino acids combine to form proteins and nucleic acids form DNA or RNA. Many other possibilities have been developed and then tested in the laboratory. The addition of RNA to a mixture of nucleic acid monomers in the presence of zinc for example will allow chains of up to forty nucleic acids in length to be copied with an error rate as low as 1 per cent.

In the five hundred million years between the cooling of the earth below the boiling point of water and the first identifiable fossils, conditions existed for the chance production of a simple small molecule that catalysed the production of copies of itself. It was not necessary to ask chance to make an organism as complex as a bacterium or even a virus, much less the hoary old wheeze of throwing the relevant parts into the air and having them accidentally combine to make a modern jet aircraft, or the production of a volume of Shakespeare by chimpanzees banging on typewriters. We are talking more realistically of the chance production of a sentence by a billion chimpanzees banging keyboards for half a billion years, that is, in excess of 10^{23} tries. There is no problem with the second law of thermodynamics as the system would be driven by energy from the sun and the cooling earth; the total entropy would still be rising, even with increasing complexity in that relatively small part of the solar system that was to become the future lakes and oceans of earth.

Once we go beyond these simple ideas of a replicating molecule, however, we run up against our present ignorance. The original molecules have long since disappeared and the early steps in the process are unknown though subject to a great deal of experimental and theoretical consideration. There is no evidence at present that life could not develop here, or for that matter over and over again on different planets, only a lack of sufficient information to show us exactly how it happened. The biologically important sugars, amino acids and nucleic acids are satisfactorily obtained in solutions from the range of simple molecules and conditions thought to be present on earth and in extraterrestrial dust clouds. Chains of nucleic acids can also be formed under conditions not unreasonably predicted to occur on the primeval earth and, given that forms of RNA are known with enzyme functions (ribozymes) and information storage functions, we are not far from a practical hypothesis for the natural development of replicating molecules. It is clear that there is no need for a complex of DNA, RNA and enzyme functions to establish a replicating system, a short RNA polymer molecule could have originally filled both the heritability role, that is, able to make copies of itself, and the enzyme role, that is, able to catalyse its own production, in the first place. The other part of the system that is needed is the equivalent of a cell membrane, and work in the laboratory exploring possible systems for the forerunner of cell membranes under primitive earth conditions is also under way.

Alternatives to the scenario described here have been proposed, including the use of compounds other than RNA as the first heritable/catalytic entity, and the possibility that the early stages of the evolution of life occurred elsewhere in the galaxy and life was transported to earth by meteors. At the very least it has now been possible to show possible routes from the primitive abiotic earth to living organisms. It is certainly not impossible.

THE EVOLUTION OF LIFE

Once we reach the stage of a simple though recognisable organism, that is a mechanism allowing the inheritance of a pattern that codes for one or more catalytic molecules maintained inside a simple cell membrane, we move on to the next stage of evolution where natural selection between competing alternative forms takes over and we are back on firm ground with observed processes in nature and testable hypotheses of relationships and ages being available. At this time, history, as particular events as opposed to the average effects of large numbers of similar events, begins to take centre stage. Up until this time we have been considering general actions that obey the laws of physics and chemistry. Very large numbers of particles, atoms and molecules have been involved and the observed outcomes have been the result of the average effects of these reactions. From this stage onwards we are dealing with historic, often once off events. We are following a particular history of a particular biosphere.

The oldest known fossil bacteria lived about 3.4 billion years ago and formations called stromatolites, believed to be formed by bacteria (at least stromatolites growing today are) are known from 3.5 billion years ago. This is about six hundred million years younger than the oldest presently known minerals on earth and about a billion years after the earth solidified. About 2.5 billion years ago photosynthetic cyanobacteria evolved. These forms used sunlight to make food with free oxygen as a byproduct. For the organisms of the day, oxygen would have been a highly toxic compound and over the next billion years its concentration in the atmosphere increased through a series of intermediate levels from nothing to close to the present 20 per cent. Because it is so reactive, it is also a potential source of energy and a range of organisms evolved in various ways, not only to protect themselves from its effects, but to use it as a resource. About 1.5 billion years ago the more complex eucaryote cells found in plants and animals evolved and about 700 million years ago multicellular organisms are found in the fossil record. A little over 500 million years ago, animals recognisably belonging to modern groups first appear in the fossil record.

The fossil record for the first 3 billion years is quite sparse, though more material is constantly being found. From 500 million years ago to the present day the fossil record is much better and detailed histories of groups and ecosystems can be studied. It must be remembered however that only a very small fraction of the species that have ever lived, much less individuals, have left a record in the rocks, and only a small portion of those found at or near the surface are available for study.

Given this situation it is still possible to find intermediate forms between many presently living major groups of organisms. For example, we have fossil and living intermediates between fish and amphibians. We have several fossil stages in the evolution of birds from their dinosaur ancestors (for example, feathered dinosaurs, birds with sets of reptilian teeth). We have many fossils showing the steps from reptiles to mammals, so many in fact that it is difficult to say when we are talking about mammal-like reptiles and primitive, reptile-like mammals. Within living mammals we have egg-laying mammals with skeletons that show reptilian characteristics. We have fossil intermediates between such extreme present-day forms as the platypus and primitive 'standard' mammals. We also have a good set of fossils showing intermediate forms between our ape-like ancestors and ourselves. While we do not have all the fossil steps in all histories we have sufficient intermediate forms to show the major steps in evolution and that evolution proceeded through intermediate forms of mixed or intermediate characteristics that were, in their own time and place, highly successful, well adapted, organisms. It is important to note that different organ systems change at different times and this is called mosaic evolution. In human ancestry upright stance and bipedal gait evolved in forms in which brain size was little altered from that of the other apes. There is no evidence of evolution through intermediate maladapted forms. When we look at human evolution we see the same pattern; the origin and accumulation of the distinctive morphological characteristics that distinguish present day humans from present day apes. As Ayala says: 'the evolutionary origin of organisms is today a scientific conclusion established with the kind of certainty attributable to such scientific concepts as the roundness of the Earth, the heliocentric motion of the planets, and the molecular composition of matter'.

ORIGIN OF HUMANKIND

Humans, of course, are not derived from any present-day ape but both groups are derived from common ancestors. Humans are most similar to chimpanzees with only a 2 per cent genetic divergence between the species. We are so similar that an example of a chimpanzee is known with the human genetic condition, Downs Syndrome, that is, three (rather than the usual two) copies of chromosome 21 that is found in both humans and chimpanzees. At present the oldest known fossil remains of the group of species to which humans belong are about 4.4 million years old. The species that these fossils belonged to is placed in the genus *Australopithecus* (or sometimes *Ardipithecus*) and is called *A. ramidus*. The next oldest known species is *Australopithecus anamensis* and fossil finds are dated between 3.9 and 4.3 million years ago. It already shows morphological evidence of an upright posture, though the size of the brain is no bigger than that of other apes. Over the next 3 million years we find a number of species placed in the genus *Australopithecus* in various parts of Africa; sometimes several species are found at the same time. The most famous of these fossils is that of 'Lucy' an adult female australopithecine about one metre tall who lived about 3.2 million years ago in what is now Ethiopia. These animals shared many morphological characteristics with us including walking upright on two legs, their hands were also similar to ours in many ways. Sets of footprints discovered in Africa and dated from 3.5 million years ago show australopithecines walked in much the same manner as we do.

About 2.5 million years ago we find a species with a slightly larger brain, 650 cubic centimetres compared to 500 cubic centimetres in the co-existing australopithecine. This species is placed in our genus and is called *Homo ergaster*. Its remains are sometimes found in association with simple stone tools. *H. ergaster* is the most primitive and oldest presently known species in our genus. It appears in the fossil record at a time of climatic transformation in Africa to drier, savannah conditions and appears when the lineages of many other groups of mammals living in the same area also show rapid evolutionary change. *H. ergaster* is followed in the fossil record by *H. erectus* which is found from about 1.8 million years ago until 0.3 million years ago. This species was the first hominid that we know of to break out of Africa and is found from Africa to Indonesia, China and Europe. *H. erectus* was larger than *H. ergaster* and the size of its brain gradually increased over time, finally reaching a size of 1200 cubic centimetres. Stone tools used by this species are more sophisticated and the wide range of climates it invaded shows that it had developed the very distinctive human habit of using cultural skills rather than morphological adaptations to survive in different environments, for example by using fire against the cold.

Ultimately *H. erectus* was replaced by modern forms of *H. sapiens*, our own species, with the oldest known fossils about 100 000 years old. Studies of the changes that have occurred in the base sequence of the human DNA of the mitochondrion (which is inherited only down the maternal line) and of genes on the Y chromosome (which are inherited down the paternal line), show that the ancestral lineages of these genes arose in Africa about 2-300 000 years ago, and that the populations to which they have belonged since have probably never been less than about 10 000 individuals. Whether these genes spread out as part of a new species, replacing an earlier species, or as genes moving through populations of an ancestral species as part of its transformation to modern *H. sapiens* is unclear at the present time. The number and relationships of the various hominid species are matters presently hotly debated, with differences of opinion between workers and the discovery of new fossils providing new insights all the time. A general overview of only the well supported aspects of human evolution has been given here.

With the arrival of modern humans we see rapid changes occurring, not related so much to morphology but to the culture of the species. In a short period we see the introduction of burial practices, the use of painting, increasingly sophisticated tools and the capacity to thrive in a wide range of ecological conditions. The tempo of evolution is increased, becoming ever more rapid, but by cultural, rather than genetic, means. It is clear that in our lineage there has been selection for the capacity to communicate and to develop complex and evolving cultures. These trends are seen in many higher primates but they reach their most extreme state in humankind.

The pattern of evolution of the australopithecines and humans is typical of that of many groups of organisms, with a general trend from generalised to divergent characteristics appearing with time. Sometimes only one species is found in an area, sometimes several differently adapted forms occur. With more detailed information, as is being steadily uncovered in the human record, the processes of change become less clear, with long periods of little change being interspersed with periods of sudden change, or perhaps the movement of species evolved elsewhere into the area where fossils are formed. The details of the mechanisms of speciation are an area of major research, with different ideas being proposed and tested by geneticists and morphologists using modern and fossil material. It is clear, however, that the evolution of humankind was not a simple linear progress of species from apes to modern humans.

4

EVOLUTION AS PROCESS

ENTITIES AND PROCESSES

The traditional way of looking at and analysing the world used by science was to view it as a series of concrete items that exist independently of their external relations (a car, a snooker ball, an atom) that interact (traffic, a snooker, air pressure), without being significantly changed by the action. The world was described in terms of these items and their interactions. Newtonian physics, with its descriptions of the motion of the planets is a fine example of the success of this approach. However, advances in scientific disciplines in the past century have challenged the idea that this is an entirely satisfactory model of reality. Studies in physics, chemistry and biology have led to an alternate view of reality in terms of events or activities. For example, we habitually see our bodies as solid, semipermanent physical presences and we frequently act as though this same body has been present for many years. The truth however is different. With each breath and each mouthful of food, within a second or so new molecules are added to our bodies; with each exhalation and each urination other molecules leave our bodies. The hand we see, which we feel has been with us for many years, has in that time changed its constitution, as the vitamins, proteins and even the calcium in its bones have turned over. In a very real sense we are not the people we were only a short time ago. Much of this exchange takes place on the scale of minutes to weeks.

Our reality is more like that of a candle flame; wax melts and is drawn up the wick where it is burned. The hot gases produced leave, the radiant heat melts more wax, but the flame continues. This dynamic state, which we call a candle flame, is not a substance but an event. It could be seen as a series of dynamic relationships in which entropy increases but complex structure is maintained, and even increases if we use our candle to light more candles, for as long as the source of energy and the relationships continue. A candle flame can cease to exist because it has run out of wax, or it can be blown out. Likewise ecological systems are apparently entities, for example the pattern of bands of plants and animals seen at different tidal heights on a rocky sea shore. It is easy to show, however, that this pattern exists only as a dynamic result of relationships. If we simply remove certain categories of organisms, for example a class of predators, the result is a different structure on the sea shore with many changes in the size and position of bands. In a similar fashion in quantum mechanics, subatomic particles can sometimes be best understood as particles and sometimes as wave forms: they are not one or the other but something different, something less easily categorised and dependent on the interactions in which they are involved.

If we begin to think of our evolution and our very selves in terms of processes, and our existence as dynamic rather than static, then we will discover ourselves and other organisms, in fact the whole cosmos, as the sum of a context, of relationships and a history.

A CONTEXT (THE ANTHROPIC PRINCIPLE)

Physicists have considered the effect of altering the observed values of the fundamental constants that are found in the laws of physics and that form part of the very basic structure of the universe. The results are surprising; the universe would not have anything like its present structure if these are altered even marginally. Similarly, a difference in the order of events in the moments after the big bang would have led to a universe in which helium rather than hydrogen was the most common element, and without hydrogen to fuel the fires of the stars, the heavier elements of which we are made would not have been produced. If the relative masses of neutrons and protons were a little more different (but still much less than 1per cent), hydrogen atoms would not be stable. If gravity was marginally greater, stars would burn out quickly and there would be no time for life to evolve.

Thus we have a set of laws underlying the universe that provide a context in which its history unfolded in an orderly way, where atoms can combine in stable ways to produce molecules, and these molecules can be assembled into the extremely complex systems we call organisms. Yet if the universe was the chance product of chaos and confusion, how was it that the laws took the form that they did take? If the combination formed at the moment of the big bang was a chance combination of all possible laws, why do we see a set of logically consistent laws (surely the chance of consistency is very low) that ultimately governed the behaviour of stars and biological molecules that were not to appear for many ages. How was it possible that these laws should have such a form that the result was a coherent universe *intelligible to creatures evolved within it*?

When we look at the laws of the universe we find a system that is logical and complete. It has an aesthetic elegance to the human mind, a harmony, a symmetry, which allows the integration of understanding to produce a set of principles that relate all things at all scales and time frames. Why is it that the language of mathematics, a system evolved in the minds of creatures that themselves have evolved, is such as to describe the laws by which these creatures are created and the system in which they live? Complexity is easier to obtain than simplicity in a complex system. The probability of the laws of nature taking on the simple, elegant, integrated pattern we observe by chance beggars the imagination.

It is worth remembering from the discussion in Chapter 1 that the laws of the universe were not easily discerned. Our superficial impressions of the earth, the universe, and the laws that govern them have often been misleading, and long and profound thought has been needed to identify a system of laws that fits the reality of the universe. These laws of the universe were not invented by humans; they were identified (even if only incompletely as yet).

It follows from all of this that life as we know it would not have been possible in a universe that had basic properties that differed even slightly from those we find. This is a general description of what is called 'the anthropic principle'. We then may well ask what is the significance of this observation?

Firstly we must be careful not to over-emphasise the low probability of the basic structure of the universe. Such a universe may be very unlikely, but its existence is a given, for we would not be here to observe it if it was not suitable to support life. The pattern of any particular set of events is very low. For example, if we throw a handful of differently shaped pebbles into the air and observe the pattern they form on the floor, the chance of this particular pattern is very low, but we would not call it significant or meaningful. If we threw them into the air and they fell forming the pattern/words 'God loves you', we would place a great deal of significance on it, though this pattern is no more likely or less likely than any other. This pattern not only meets the requirements of the laws of nature but clearly includes symbols placed in a suitable combination and a deeper meaning is also present behind the words formed. Thus the elegance, simpl... ...y and integrity of the observed pattern of the laws of the universe needs our interpretation, like the pattern in the pebbles, both as a symbolic language and as a message. While there is no irresistible physical argument that forces us to recognise the reality of the beauty, symmetry and life-giving character of the universe, we are, however, free to do so.

However we interpret this observation, the order, elegant structure and the orderly unfolding of the universe provides the context in which we live, in which our theology must be developed and within which it must find meaning.

RELATIONSHIPS

The idea of relationships has become critical in several branches of science during the past century. In physics the rise of quantum theory has led to unexpected, and, to most people, strange insights into nature. One product of this research has been the insight of the connectedness of the act of observation and the state of a particle. It seems that matched pairs of particles (for example photons) that result from some action, for example the decay of an atom, have opposite states, such as spin. Which particle has which state however is indeterminate until it is measured. At the moment of measurement, the situation for one particle, which until then was a probability, becomes determined. At the same instant, however, the state of the second particle in the pair is also fixed, no matter how far away it is. This 'second' action, though in fact it is actually an aspect of the 'first' action, not something separate, happens instantly, faster than the speed of light, that is impossibly fast if our present physical theories are correct. The theory that underlies and predicts this observed result has been taken to imply that an innate instantaneous connection exists between all particles in the universe and that the state of the universe is at least in part dependent on the act of observation or measurement of its state.

At higher levels of complexity, for example in ecological systems, we have seen that relationships determine the nature of the existence of such systems. It is more than a simple matter of cause and effect, their actual existence can only be found in terms of relationships. Similarly, humans interpret the world, including themselves, and shape their actions in the context of their experiences. These experiences are in fact of relationships. As a consequence, our capacities as individuals largely have developed through our relationships with others and the physical world, and have been expressed in the context of those around us now or in the past. In a very real sense we only exist in and through our relationships with others. A human then is a physical entity subject to the relationships identified by quantum physics, a biological entity, as reflected in our biological relationships, and a human being as a consequence of our physical, psychological and spiritual relationships.

A HISTORY (CHANCE, NECESSITY AND CHOICE)

It seems likely that all present living forms on earth are derived from a single cell that lived more than 3.5 million years ago. There are two sets of observations underlying this conclusion. Firstly, this date lies between the time when the earth cooled to the point where life was supportable and the time of the oldest known fossils. Secondly, a single origin of life is assumed because the same general forms of DNA and information coding processes are used in all bacteria and higher organisms, in combination with the same set of amino acids, though a wide range of other compounds and mechanisms are quite feasible. In the scenario described in the previous chapter, it would be expected that a much wider range of options would have been explored in the early stages of evolution, however only one lineage survived. As a consequence, we are ultimately related to all living organisms and particular events – namely the survival of particular organisms – have left their mark on all their progeny. We are here dealing with the particular history of a particular planet.

The process of evolution depends on a creative tension between stability and change. If the order of nucleic acids in a particular DNA sequence in a species was rapidly changing, for example through a high frequency of copying errors at replication or through the effects of mutagens, there could be no stable forms, no stable species adapted to their habitats in the context of other equally well adapted and stable species. On the other hand, if there was perfect stability, with no change, then no new forms could arise and all living forms would be identical with the first. Consequently, while the mechanism of copying and error correction are very good, they still allow chance errors to creep in at the rate of one in a million or so. Most of these mutations are deleterious as they cause the proteins they code for to be either nonfunctional or less well adapted than the normal form. This is not surprising as it would be expected that the form commonly found would be well adapted, through the refining effects of millions of generations of selection, and the chances of a new mutant being better adapted are correspondingly low. Mutations do not occur in response to need and are not produced to meet a specific need. There is simply the continuous production of a pool of variants for each gene.

Occasionally, of course, an altered sequence is better adapted than the usual form and will replace it over time. Perhaps more commonly, as conditions change the opportunity arises for previous less well adapted forms to become the better adapted form. For example, the genes that provide resistance to the insecticide DDT in insects have perfectly ordinary functions, and the mutant forms that provide resistance are less well adapted to these roles. For millions of years these forms (along with thousands of other forms) kept appearing as chance mutants in the many different evolutionary lineages of insects. They were removed by selection through the death or reduced breeding efficiency of the individuals containing them. Finally, we humans began a massive campaign of using DDT to destroy malaria mosquitoes and agricultural pest species across the world. Within a few years DDT-resistant forms steadily increased in frequency in species after species. That is, in the changed conditions, different forms of proteins from those previously common were at an advantage. In situations where we have stopped using DDT, the balance moves back again and the frequency of DDT-resistant genes in the population slowly goes down again. Nature is profligate in the production of individuals and variation, with the birth and death rates needed to maintain this capacity also prodigious.

Another instance of these phenomena occurs when a gene changes its function. For example, when a gene is duplicated and selection specialises each copy for a different role, we find increased rates of change in the sequence until it is well adapted to its new role. Such a result is seen in the evolution of haemoglobin, where, in an ancestral species to the mammals, the same form of haemoglobin was used in both foetal and post-natal respiration. Duplication of this gene so that each individual organism had two functional forms of the gene, allowed one copy to better adapt to the needs of the foetus while the other was adapted for respiration after birth. In the several million generations following the duplication, the sequence of the foetal haemoglobin changed rapidly but then the rate of change slowed to that seen previously. Thus, after the gene becomes well adapted to its function the rate of acceptance (NB not production) of new mutants steadily drops to the background rate as the likelihood of finding a better adapted mutant form decreases.

It is noteworthy that, when considering the production of variation, we must also realise that these mutation events are truly chance; if this specific universe could be re-run, the results would not be the same. Quantum physics thus proposes that some categories of events are truly random and are believed to be inherently unpredictable, and not just unpredictable because we do not know the mechanism of occurrence. An atom of a radioactive element will break down at a truly indeterminate moment. At its deepest level the universe is not mechanistic and this is also manifested in the production of genetic mutations.

In summary, then, we find that mutations occur and these mutations are not related to need: firstly because the mechanisms of production are chance, and secondly because there is no way of identifying in advance what might be needed. There are some examples of genes where the need for change itself has been selected, for example in the immunoglobulins, but these are a small, highly selective set of cases. Thus the particular mutations that occur are due to chance, though the rate of their production has probably been established as an adaptive trait in combination with birth and death rates. The rate of acceptance, or survival of a mutation, depends on the probability of finding a better adapted form of the gene than those forms already present in the population. It can be seen then that variation can occur without conflicting with the second law of thermodynamics. Biological innovation is simply the sum of such events and therefore also not contrary to the second law. In so far as the second law of thermodynamics deals with change, it is with the thermodynamics of the substitution of one nucleic acid for another in a DNA sequence, the fact that this base change leads to what we see as a highly innovative change in function in the resulting organ system is not subject to the second law. Innovation is an anthropomorphism we apply to changes observed in nature, not a physical/biological one.

The long term survival of mutant forms of genes depends on their relative 'fitness', that is their capacity to leave more survivors than alternate forms. Most mutants are quite unfit and are quickly removed from the population by selection (these are lethal mutants, and are removed by what is known as 'hard' selection). In such cases a death intervenes to remove the newly created mutant form and there is one death for each case. Even in humans more than half of all fertilised eggs die before they reach the implantation stage. A few versions of each gene, however, have survival characteristics that are not greatly different from the usual form and can survive, at least for some generations. The critical issue for these non-lethal mutant forms is the intensity of competition between individuals. Every individual in every species has these slightly less suitable forms for many genes. Each individual has a different number and combination of them. All of such individuals in a population could survive quite satisfactorily if there were sufficient places for them all as adults, but there are not, and competition occurs between them for a place. The outcome of this competition depends on the overall relative survival characteristics of each individual. The necessary death rate to leave only sufficient survivors to fill the available places in the breeding population may be reached through differential survival of variants at other genes, allowing individuals carrying the variant gene we originally considered to survive into the next generation. This situation is called 'soft' selection. That is, the simple hard selection model of death or survival does not occur for this class of variants; it is a matter of 'more or less fit' rather than 'fit or unfit'. The amount of such variation carried in a species can be seen for example when it is 'uncovered' through artificial selection in the production of domestic breeds. Consider the range of variations in size, form and behaviour found in dogs, all of which were present in the population of wolf-like precursors. In nature, changed conditions allow the adjustment of phenotype by changes in the frequency of these variant forms of genes to occur as rapidly as it does in the production of domestic breeds, that is, within less than a hundred generations, and nature has millions of generations to explore options and refine products. Another key effect of soft selection is that the death of a single, less fit, individual removes from the population less suitable mutant forms of many genes at the same time.

It follows from these considerations that no species or population is perfectly adapted, it is adapted well enough to survive in its present context, that is all. The world is always changing; it is only 15 000 years since the end of the last ice age; insecticides and antibiotics produced by humans have only been around a short time. The process of adaptation must continually be redone as conditions change. 'Micro-evolution' of this kind is easily observed both in nature and in the laboratory. It is easier to conceive of a species then as a highly responsive process rather than a static entity. In biology, a species is defined as a group of actually or potentially interbreeding individuals; the definition is of a process, not an entity.

Species, however, sometimes undergo directional, systematic change that can lead to the origin of new species, for example in response to changed conditions. The production of new species is normally (though not necessarily) a much slower process than the process of micro-evolution we have been considering to this point. Speciation usually requires the accretion of so many changes in the genotype that ultimately interbreeding is impossible between the 'old' species and the 'new', due to mismatching of the developmental instructions. Speciation can occur quickly, however, and new species have been produced in the laboratory. An area of intense study and debate in biology is the process of speciation. While there is no doubt as to its reality, the mechanisms, be they one or many, are still being studied. It is unclear whether speciation occurs slowly and steadily or whether it is a relatively speedy process interspersed in each lineage with long periods of morphological stability. I suspect both processes are found, depending on the circumstances.

The evolutionary process then, leads to species that have integrated suites of inherited morphological and behavioural characteristics adequately adapted to the conditions in which they find themselves. It is this exquisite pattern of adaptation that led to the defence of theism used in the period before Darwin published *The Origin of Species*. The argument is based on an analogy. It suggests that one would find it difficult to believe, on examining a finely made watch, that there was no watchmaker; that the watch had, as it were, made itself. In a similar fashion when one looks at the beautiful and finely made mechanisms of plants and animals, it is difficult to believe that these species had made them themselves. There must needs be a watchmaker, and that is God. However, the concept of evolution by mutation and selection provides just such a mechanism for species to 'make themselves'.

The anthropic principle considered previously could be seen as again raising the watchmaker argument at a further remove in time. The laws of the universe are a beautifully made system and need a maker. The differences, however, are significant; we are not here talking of a 'God of the gaps', an entity conveniently plugged into holes in our present scientific understanding but of a creator at the beginning of time.

In this context an example may be helpful. If we look at the human eye we find a wonderfully adapted instrument consisting of many parts that must all work in harmony. Each part must be of the right shape and size, and in the right position for the eye to be functional. Remove any one of the parts and the result is non-functional. This has been the example used of the 'watch' that needs a watchmaker. It is claimed that it is not possible to build an eye by evolution because, until all the pieces are in place, it is not of any use. For those that follow this line of argument however, the answer was demonstrated over a hundred years ago. For mutation and selection to work, it is essential that each intermediate step in the evolution of this complex eye must be itself adaptive. While the intermediate forms leading to the human eye cannot be studied from the fossil record (because they are made of soft tissues and therefore not preserved), we can find all the intermediate forms in presently living species, from simple photosensitive cells (allowing the detection of changes in light intensity), through cup-shaped complexes with photosensitive cells inside and opaque sides and bottom (allowing the direction of light to be detected), to the addition of a simple lens (that allows movement but not an image to be detected), to a focused image of increasing quality (allowing shape and distance to be detected). In each case the organ meets the need of the species possessing it and is, in context, adaptive. The argument that non-adaptive intermediate forms are needed in the evolution of the human eye is false. This example also shows intermediate stages in the morphological evolution of an organ system can be easily found in nature and that the production of innovations and novel systems are possible, solely by the action of evolutionary mechanisms. Systems as complex as the eye are also found at the biochemical level. Many of these are common to all presently living species and so must have evolved more than 3 billion years ago. At this stage it has not been possible to show the steps in the evolution of this biochemical complexity, as we have done for morphological complexity, as none of these early forms survive. However, the proteins of early ancestral forms would not have been as functionally specific as present day proteins. Even today most enzymes will catalyse reactions using many different substrates.

Recently the evolution of the eye from simple precursor tissues has been modelled using a computer program that mimics the selective pressures and modes of inheritance of the respective tissues. It was found that a complex eye could evolve in as few as 10 000 generations, compared to the two billion available.

The reverse proposal however, that a watchmaker could have made the watch (for example, an eye), is not supported. The human eye is not as it would be if it was designed by a master engineer. There are fundamental design flaws in the eye that would not occur if the eye was produced from scratch specifically for human use. The nerves that run from each photosensitive cell to the brain are attached to the front of each cell. They combine to form a nerve fibre that passes out through a hole in the retina and so backwards to the brain. As a consequence, firstly, the nerve fibre is an obstruction affecting the quality of the image produced and, secondly, a blind spot appears in the field of vision where the fibre passes through the retina. Any sensible designer would have simply avoided these problems by running the nerves off the backs of the photosensitive cells. There is no innate problem in doing this, as it is the pattern used in the independently evolved (and equally complex) squid eye. Many such design limitations are known.

Such imperfections in design arise during evolution because the process can only work from the present state of a characteristic, the range of adaptive possibilities available at the time and the needs of the organism at the time (not the needs of some future daughter species). In a distant precursor species to humans in which a ball of photosensitive cells served, simply and adaptively, to detect changes in light intensity, it did not matter which side of the cell the nerve came off. It was only much later in the evolutionary process that it became a problem, and then it was not possible to 'back up and start again'. In evolutionary terms one can only work with present opportunities and confront present challenges. The problems of the design flaw in the human eye are, of course, covered as adaptively as possible. For example, the nerve fibre is partially transparent and the blind spot is off centre so that the two eyes are 'blind' in different places and the integrated image developed by the brain from the signal from the two eyes covers the problem. Nonetheless, it is a flaw in the basic design that would not have happened if the human eye was the intended product of a design process from the first.

Thus the natural world is not perfect; not only that, it is not a degenerate derivative of a perfect creation. It is simply the product of living and exploiting the opportunities available at each moment.

The explanatory power of the concept of selection as the mechanism for identifying the best adapted form from amongst the options available has been extended in the past two decades to include the concept of 'inclusive fitness'. The traditional view of survival of the fittest has been transformed to that of 'leaving the most copies of your genes in the next generation'. The critical effect of this change is to reduce the necessity of the individual to breed. There are other ways to leave genes in the gene pool, particularly by aiding your close relatives to breed; for example, by assisting your parents or siblings to breed successfully you may leave more copies of your genes in the next generation than would result from breeding yourself. The classic example of this behaviour is seen in the Hymenoptera where the complex patterns of social behaviour seen in ants, bees and wasps, have arisen perhaps sixteen different times in different lineages, while in all the rest (75 per cent) of the insects it has appeared only once. The reason for this odd pattern can be explained on the basis of the genetic system of these animals. Unlike the normal system found in the rest of the insects and other animals, female ants and bees are genetically more similar to their sisters than they are to their own young. Consequently there are stronger genetic advantages in these groups to giving support to your mother's breeding than in having young yourself. By using this model of inclusive fitness it becomes a great deal easier to understand the evolution of behaviours like altruism in species where communities consist of close relatives. In such a situation apparently altruistic behaviour can have an adaptive basis, thereby answering the real challenge as to why any organism should die for another (that is, kin selection). This situation is thought to have occurred in human evolution where social groups consisted of related individuals, thereby providing the context and driving force for the evolution of social behaviour in our ancestors.

The process might be seen as analogous to the difference between selecting the suite of attributes needed in an athlete able to compete and win in a single sculls race and selection for the attributes needed in the members of a winning eight man rowing team (that is, group selection). While the latter includes the former, a great deal more is required to produce a winning team rather than a winning individual. This is similar to the biological attributes needed to allow humans to live and thrive in teams (or clans). If these attributes are at least in part inherited, then selection can quickly favour the survival of such clans over others. The acquisition of many human attributes, for example, language, cultural/technical skills and morality, all require the individual to have the mental capacity to learn these skills. There is, however, a marked difference between capacity and content. A young child can learn any language because it has the capacity, the language learned depends on context. A similar situation exists for other attributes. It has been argued that the capacity for young adults to accept and commit to the clan's religious faith (and hence the clan) is equally a capacity with survival value. Again, however, it is a long step from capacity to content, though developing 'the ears to hear' in the spiritual sense would be a necessary evolutionary step into the fullness of the human spiritual condition.

Thus, the essential selfishness that would be predicted from a narrow view of survival of the fittest can be transformed through kin and/or group selection, allowing the development of a range of genetically based behaviours that have adaptive social consequences.

In the traditional mechanistic model of the universe, then, we would see the explanation of evolutionary processes in terms of two mechanistic forces: the chance production of variation, which is the product of physical processes, and necessity, that is, the challenge of death or survival in a hard selection model. The process, however, is not quite so easily described and we do not live in such a limited, mechanistic world.

We find instead in nature a steadily increasing range of possibilities opening up with the increasing complexity of life. From the protozoans onwards the behavioural interactions between organisms and their environment, including other organisms of the same or different species, become more and more critical to survival. In animals of increasing complexity, the roles of exploratory behaviour and learned responses become more and more important. A key attribute of advanced learned behaviour is the ability to make choices between options on the basis of experience. In the birds and mammals there are many cases of significant changes in behaviour on the basis of insightful interpretation of experience. Even in the insects we see community decisions made by consensus (for example, in the choice of a new hive site for the swarm by bees). These choices change the conditions in which a population lives thereby placing it in a new context, and so open to selection under different conditions. In the primate lineage we find these choices leading to conditions favouring the selection of ever more complex behaviour. Ultimately the genetic constitution, expressed in the form of the physical body of each individual, can support, in many primate species including humans, the development of distinct learned and evolving cultures requiring no changes in the genes (for example, the different and evolving cultures that have been described in Japanese macaques). In the human lineage the process driven by inclusive fitness has gone even further, ultimately leading to the suites of inherited characteristics underlying the development of speech, morality and the other distinguishing attributes of the human condition.

The interaction between chance, choice and selection therefore underlies the evolution of the human condition, and to a significant extent that of all higher animals. We are not simply the products of a system that can be reduced to blind chance and implacable necessity; choice also plays a part; we, in nature, are at least partly what we have chosen to be. We are the product of a process, a process in which life has been a participant, not just an object on which the forces of nature have had their way.

DEATH

If the laws undergirding the universe were such that no living creature could exist, then there would be no sickness or suffering, there would be no injustice. However, the universe clearly does allow the possibility, perhaps even the certainty, of life and, consequently, we are left with the unpalatable fact that its Creator was and is willing to accept suffering and death as essential ingredients in this universe. Many would say that the price paid for life is too great, and we will need to look further at this issue later in the book. For the moment we may ask where, then, does death fit into the material scheme of things?

Death is a critical component of the process of evolution and enters our calculations in several ways. Firstly, the reality of death is an essential part of the nature of life. In so far as life is a process and increasing entropy is the unavoidable consequence of a missed balance, we find that death is the universal fate of all creatures. It is a chastening thought that if we were able to remove death from the universe we would stop any chance of change, we would remove any chance of a new generation taking its place of leadership and responsibility, we would hasten the death of our planet through overpopulation.

Secondly, all animals and other organisms require the deaths of other individuals if they are to survive. Carnivores have jaws, teeth and digestive systems designed for eating meat; they cannot live on a diet of grass. Their instinctive behaviour patterns are also those of hunters, not of the hunted. They must kill to live; they must kill to be truly themselves. Even those organisms that do not kill are dependent on nutrients that are recycled from the dead bodies of other plants and animals. The available nutrient pool is relatively small and would be very speedily used up if it was not replenished. Much of what many view as evil is inseparable from life.

Thirdly, different survival rates for differently adapted individuals leads to better adapted populations, and differential survival, of necessity, requires deaths. The necessary condition for the emergence of more complex and better adapted forms is the deaths of less well adapted forms. Thus individuals die and species become extinct. In effect, we only exist because of the death of many species. In so far as those things that make us truly human are the result of evolution, we find that death, competition, struggle and suffering are essential elements of the process by which beauty, consciousness and understanding are made explicit. For the theist, creation is intended to bring forth these values, they are no accident, any more than is increasing entropy, they are implicit in the whole, but death is the necessary cost of their development in this universe. 'No pain, no gain' is a law of nature.

Fourthly, old age, that is, senescence, follows as an evolutionary consequence of the reality of death. Senescence might be seen as the steady decrease in our capacity to manage the processes of life, that is, our biological control mechanisms and repair mechanisms become less able to respond quickly and accurately. The origin of this effect can easily be understood in evolutionary terms. We must remember that we do not die of 'old age'; we die because we cannot respond to biological challenges adequately, whether it be because of slowing response times when driving a car or our ability to beat off an attack from a disease organism. In effect, it is as if the candle flame is disturbed and our ability to recover a proper dynamic balance is reduced; will the candle of life continue to burn, or will it go out?

The evolution of old age starts from the production of mutants that increase survival or reproductive capacity in younger organisms but at the cost of reduced reproductive capacity or survival at a later age. We must remember that evolution is selecting total reproductive success over a lifetime. Given that death steadily reduces the number of survivors in a group of organisms as time passes, it can be easily shown that the advantages of increased reproduction at an early age can outweigh reproduction for a longer period (because we may not live long enough to breed later). As we would continue to die from natural causes, even if we were potentially immortal, there are definite advantages for a population that breeds early, before too many have died, or that reduces death from natural causes at an early age, if necessary at the expense of more serious deterioration at a later age. It is no good potentially being able to live forever if you die because of slow reflexes at eighteen. On the other hand, in species like our own that have a long growth period, selection will also work to ensure offspring will reach independence (at least) before parents die. The consequence of all this is the steady accretion of alternate forms of genes leading to better survival and reproduction in the young, even if these same forms have deleterious effects on long term survival. Old age is not due to the effects of a single gene, then, but to the steadily increasing effects of many genes. It is not a condition that can be easily treated.

Biological death and the suffering that goes with death, then, are necessary parts of the structural reality of this universe. Any adequate theology will need to take account of this reality.

PROGRESS

When we look at nature we can easily imagine a driving imperative for increased complexity. As we move forward in time over the ages we find evidences of increasingly complex organisms. This increasing complexity continues until the development of the human brain occurs, the most complex structure in the known universe. It is natural for us as the recipients of such brains to see this increasing complexity as 'progress', specifically progress towards us! While the idea of progress is an issue of interest in many contexts, we must ask ourselves about the relationship between progress and evolutionary increases in complexity.

Firstly, we can say that there is evidence of increased complexity over time. However, was this directed, an imperative of nature? Evolutionary theory would at one level say no, it is not. What we find is an imperative for increased inclusive fitness. While this may mean increasingly complex nervous systems in some lineages, for example, some primates, this is by no means the usual pattern of improving fitness. We see increasing fitness in many parasites, for example, leading to steadily less complex nervous systems (though nonetheless well-adapted forms). An examination of any group of organisms shows a range of forms adapted to take advantage of every opportunity. These suites of adaptations draw out a sense of amazement in zoologists. It is truly wonderful how 'ingenious' these adaptations of standard forms to special conditions can be! The overwhelming sense is one of exploration of possibilities. If there is a way to exploit a niche some lineage will find it and develop its capacities to survive there. The sense, then, is not one of progress or even increasing complexity, but one of exploration.

Given then that the evolutionary process began with simple organisms, it would be expected that the exploration of possibilities would include increasing complexity, though not necessarily a drive towards complexity. After all, 3.5 billion years to reach the complexity of humankind is not exactly rushing, but then speed is simply not a relevant factor for a Creator outside time! The development of what we consider human attributes and capacities could have arisen in any of a wide range of evolutionary lineages, for example the octopuses or dinosaurs. Whatever we mean by progress in a biological sense is not restricted to the necessary development of our species, or for that matter to organisms on this planet.

At the theological level, the model of evolution described in this chapter includes some critical factors that are implicit in the very nature of the universe and they provide us with some challenges we must face. Clearly, the whole concept of the universe and ourselves as process needs to be considered, as does the relationship between chance and free will (choice). The overwhelming profligacy of the evolutionary process, with the enormous number of deaths needed to run it and the huge amount of pain that results from the combination of mutation, selection and choice, must be addressed in any honest theological discussion.

The imperfections in design that have resulted from the very nature of the process also cannot be ignored. Finally, the partially blind and partially deliberate exploration of possibilities we see, rather than some simple process of upward progress must challenge any simple view that would see us as the intended 'Gods of creation'.

5

CREATION AND EVOLUTION

The natural world, its history and processes, described in the previous two chapters can now be used as data for theological consideration. It is clear that the world view painstakingly developed by science over the past two hundred years changes significantly the context in which we must make sense of faith. In this chapter we will reappraise some aspects of the idea of creation and in the following chapters look at Christian ideas of redemption in the light of the insights of science and then of our role in creation. To begin our analysis we need to remind ourselves of the point made in Chapter 2. Modern natural theology is based on the assumption that the world is the result of the creative acts of God; it reflects the mind of God. As a consequence we have information on the mode of action and intent of God and of our place in this intent. Clearly it is not a replacement for revealed insights but our interpretation and integration of both sets of experiences must lead to a consistent theology.

Before attempting this reappraisal, however, we need to question whether, in the changed circumstances, Christianity is still a suitable frame of reference. Certainly many in our society do not believe so. The major Western alternate presented is materialism. This is a system of belief that assumes that nothing exists beyond the material universe. This world view assumes that there is no world of the mind or spirit beyond the material fact of the human brain. These so-called 'extra realms' are simply side-effects, epiphenomena, of brain function. It is also taken, as a corollary, that there is no God. It is important to realise that this world view is a faith statement just as much as that made by a theist in saying that there are realities beyond the obvious factual existence of the directly observable material world.

It is a common practice of some scientists who happen to be materialists to equate materialism with science; to speak as if they are one and the same thing. They can then claim the success of science as proof of the validity of materialism. This is of course nonsense. There is no such necessary relationship and such a position is a form of the religious belief we called scientism in Chapter 1. In making this claim they have also done a great disservice to science, as we shall see later.

In Chapter 1 the use of Occam's razor in science was described. That is, the hypothesis that makes the fewest assumptions is to be preferred, because it is the most explanatory and the least *ad hoc*. Materialists have claimed that their philosophical system is the simplest and makes the least assumptions of any faith, and therefore should be accepted by any rational person. Materialism, however, is an oversimplification. Ward describes it as 'exclusive' simplicity, that is, it denies the real existence of all that it cannot explain. In Chapter 4 we saw that the material world is supported by a coherent and elegant set of laws (though, if the initiation of the universe was an accidental, spontaneous event, then chaotic structures are more probable), we have seen examples of new sets of functional laws and attributes arising with increasing complexity (and that these were implicit in and consistent with, the structure of the cosmos, though not reducible to the laws of physics). As humans we speak of such things as mind, awareness, purpose, beauty and value that are defined out of existence in materialist philosophy (that is, they have no real existence as they are simply epiphenomena of the brain). The ultimate epiphenomenon is rationality, but if the world is as they claim even this disappears and they are left attempting to make explanations using mental mechanisms and values (for example 'true' or 'false') that do not exist. Questions such as 'Why does the material universe exist?' are not allowed as they are meaningless (as there is no such thing as meaning) and valueless (as there is no such thing as value)! This is not to say that materialist scientists are not committed to, and act as if, these things exist. They are at least as committed to truth and beauty as scientists who are Christians. It is just that the intellectual foundations of such commitments are rather doubtful.

Ward contrasts this position with what he calls 'inclusive' simplicity. This recognises the reality of consistent law, increasing complexity and value, and proposes theism, the existence of one God who is, in Ward's words, 'the source of all being and value, of freedom and necessity, of unity and diversity, of matter and consciousness' as the basic assumption and ultimate reality. Such a general principle has the simplicity and elegance we have come to expect from reality as reflected in science, and it accepts and integrates all the complexity of life into a single unity. It provides a unifying context that accepts the reality of consciousness and a sense of moral obligation, of beauty, culture and value along with freedom, purpose, chance and necessity as realities that can be integrated into a coherent world view. As our knowledge of the cosmos grows we are constantly in danger of being guilty of the accusation made by Phillips, 'Your God is too small'!

The consistency of the hypothesis of God with our experience of the cosmos cannot of course be used to prove the reality of God.

It is, however, what we would expect of a God of the kind proposed, and the absence of such consistency would count heavily against such a proposal. It is also clear that we could not deduce the God we find in Christian revelation from our study of the process of the world, but they must be consistent. Our concept of God must fit the facts of the cosmos as well as the facts of revelation.

It is interesting that science, with its need for coherence, consistency and rationality, was originally developed in just such a religious context and there does not seem to be any good reason to discard it.

FREEDOM, PURPOSE, CHANCE AND NECESSITY

The physical and mental worlds we see are constructed from molecules and atoms that themselves are constructed of particles from the world of quantum physics. However, as these particles are combined to form systems of greater and greater complexity we see the expression of attributes that are not observable and not even predictable in the simpler precursor systems. A knowledge of chemistry will never lead one to the laws of genetics, much less to psychology. These novel expressions of the potential of the universe, which were described in Chapter 4, show us a cosmic system that has the potential to explore possibilities.

The work of quantum physicists leads us to the view that if the cosmic system could be run again from exactly the same starting conditions, we would not obtain the same outcomes. Further, the concept of chaos has become of critical importance. The responses of the world to change are not linear; for example, in ecological systems we find that the size of the response to altered conditions is not proportional to the size of the alteration. A minute change in the value of a variable can make radical differences to the way a system functions, while, in another situation, quite large changes may make little difference. It is, for example, for this reason that predicting weather conditions for a particular day well in advance of the event is impossible. There is a freedom implicit in the structure of the universe that defies our capacity to predict outcomes. This defiance is not simply a matter of needing more accurate measurements. There are classes of events that cannot ever be measured to the accuracy needed. It is like trying to measure the exact value of $\sqrt{2}$ or π. Further, the accidental intersection of independent causal chains (for example, the world of the dinosaurs and the causes of the Cretaceous terminal event which led to the destruction of that world and opened the door to the evolution of the world of mammals) is independent and the consequences of their intersection essentially unpredictable. The universe at heart is open to the future. The consequences of this openness are both good and bad, and we will consider this further in the next chapter.

At the same time, if we look at other scales of events we find well ordered systems capable of high predictability. For example, again with weather prediction, the El Niño system allows average rainfall and frost rates to be predicted with good accuracy some months in advance in southern Australia, though not the conditions on any particular day.

Chaos and order are both attributes of this universe and they are complementary and both are implicit in its structure. While such a mixture of randomness and orderliness has been interpreted by some as demonstrating the meaninglessness of the cosmos, it can also be seen as providing the mechanisms needed for the free exploration of possibilities. The effects of quantum uncertainty and chaotic processes provide the variation needed for order to accept or reject. The combination provides a system of great fruitfulness where the possibilities implicit in nature can be explored. Given that these attributes are implicit in the structure of the universe they must be considered as part of its Creator's intent in creation. The universe and the creatures it developed are expected to be independent, to explore and to choose. Further there is an openness in the system that means the future cannot be predicted, as it is ultimately unpredictable.

What, then, are we to make of the idea of the foreknowledge of God? If the entire time context is part of the nexus of laws and energy that constitute the universe as we saw in Chapter 3, then God, as creator, is 'beyond time' as St Augustine so elegantly showed. The entire history of the universe from big bang to its ending is perceived by God, from the 'outside', as a single entity with all events and choices known. When viewed internally from any point within time, however, it is indeterminate, its future still to be made. Freedom or predeterminism are a matter of one's viewpoint from inside or outside the cosmos.

However the universe ends, it is difficult to see that any purpose intended in its creation will be fulfilled simply by reaching its end. It is hard to see either the big crunch or the inexorable rise in entropy until nothing more can happen, as of value in themselves. However, if we contemplate the stages of evolution; for example, the ignition of the first star, the life of the world of dinosaurs, or the satisfaction of listening to beautiful music well played, we can recognise that worthwhile states have been present so continuously that we must see the process of the universe as a whole as worthwhile. The universe is purposive in that a single, integrated and coherent system of elegant laws and energy can bring forth an almost infinite number of states, entities and situations of great value, but progress to some ultimate destination is not its purpose.

We see, then, that for between 12 and 15 billion years the cosmos has produced situations of great value in the absence of humans. The value and worth of the universe does not depend on our presence. Further, examination of the evolutionary process shows that it does not necessarily contain humans as we understand them, though it does contain the capacity to produce creatures that can know and respond to God. There is no evidence of progress, only of a unique history, that is, the irrevocable passage of time and the interconnectedness of events between the past and the present. We certainly see the exploration of possibilities, including the possibility of increased complexity, though evolution is not restricted to this for we also find, as we saw in Chapter 4, increasing simplification. Not only do we not see progress, we find that evolution is a makeshift, ramshackle process 'satisfied' to make do with an adequate, opportunistic creature sufficiently adaptive, but filled with design limitations as a consequence of its history. While we should take ourselves seriously, for we see in Christ that God does, we should not take ourselves too seriously, nor see ourselves as either the centre, or our evolution as the only purpose, of the created cosmos.

Since time immemorial humankind has looked at the cosmos for clues as to the meaning of life and for ways to manage lives. It is surprising then that as science gave us a sound basis of information on which to draw for theological reflection, many have taken the view that, far from being a repository of meaning, the world is without point or purpose. This view is held even though science takes as an assumption that the world is ordered in its fundamental structure: an assumption that has been sustained by two hundred years of research. The attitude of materialist scientists can be summarised in the claim of Stephen Weinberg that the more comprehensible the universe has become, the more pointless it seems. A sense of futility and despair stalks the high technology world that we Westerners have built and inhabit.

We find clues to the cause of this move from meaning to pointlessness in the life of Charles Darwin. As a sensitive spirit he found the cold, ruthless and profligate levels of death and cruelty in the natural world in general, and particularly in his family experience, contrary to any gentle vision of a creator God. He was quick to recognise the byways that nature took, and saw there was no consistent systematic drive towards a set of goals. There was no evidence of any effort to take the shortest route to presumed goals (for example, humankind). The sheer indifference of nature begged an explanation if there was to be a purposeful, not to say loving intent, behind the forces of nature.

Clearly a God based on the watchmaker model popular in his youth was not going to be adequate. Even given the reality of these observations, they are not all the picture. Integrated and functioning communities, marvellous and beautiful structures and ingenious solutions to the challenges of existence are to be found in the world. The book of nature is ambiguous to say the least.

Clearly when scientists, as ordinary human beings, respond to these observations by finding or not finding meaning, they are responding simply as humans. Science of itself does not provide meaning, it must be interpreted by faith, whether it be materialism, Christianity or some other. It is true however that any view we may hold on such matters must be consistent with the findings of science at the deepest level. We could go further and say that any responsible theological position must make sense of these observations and fill them with meaning. We must ask ourselves: is it possible to build a world view based on hope rather than pessimism in such a universe?

As we have seen in the previous chapter any attempt to include God in the process of evolution as the 'intelligent designer' or the 'master engineer', is doomed to failure. The world shows no evidence of such activity and much evidence to the contrary. There is in fact no theological reason to defend the idea of a God who acts as master puppeteer in cosmic history. Such a view of God is not the Christian view in any case, so we can consign it to the dustbin without a qualm. The New Testament view of God is as self-emptying love, where love may be defined as 'enabling the beloved'. We are to understand that we have a living relationship with a compassionate God who pours out love on creation in the most vulnerable way, by partaking of its pain as a human being. The biblical record is one of a long relationship of promise without compulsion, one in which the future is always open, one in which a crucified and risen 'saviour' offers us, but does not require of us, the choice of accompanying Him into the future. The Christian view then is one of self-humbling love offered in such a fashion and in such a context that new and different futures beckon us on, as we will see in the next chapter. This freedom is perhaps like a parent allowing a teenage son to take the family car out on a Saturday night, or a mother allowing the resources of her body to be used by another individual growing in her womb. These things are given in love and with the intent of providing the opportunity for increased freedom, despite the known risks involved.

Can we then apply this vision of God to the evidences of science, and can we come to an understanding that is not based on futility? Such a view is no more (or less) a faith statement than that which views the cosmos from a materialistic perspective. As we said earlier, it already makes more sense of such science-driven perspectives as the intelligibility of the world and the value of seeking the truth about reality than does materialism. Our view would see the world standing continuously in the sustaining love and promise of God with the world's existence and the existence of its laws sustained for as long as God chooses. The creative mechanism used by God in this work of creative love is the freedom provided by the evolutionary process.

If we accept the proposal that love is one of the key attributes of God, then we must ask ourselves questions as to how this love would be expressed. Love consists of a special kind of relationship and it is a process, not an entity. True lovers (whether they be men and women, God and human, human and nature) evoke and promote the other's identity. As a consequence, for humans love will lead us towards wholeness and maturity as individuals. We are strengthened by the relationship, but not subsumed by it. Real love gladly gives up selfishness and suspicion and transforms the distortions of our personalities that result from our previously self-centred perspectives. More generally then, we can see love as 'enabling the beloved' and it follows that God cannot/will not coerce creation, given that providing a partner for love was the purpose of creation. We would expect such love not to overwhelm its creation nor to show itself in such power as to annihilate it. The consequence of this is that the cosmos, seen from the inside, should not show evidences of direction but of the gift of opportunity. Creation then is not the stepwise unfolding of a previously developed detailed divine plan but the provision of the freedom and opportunity for creation to explore, become and share. Love requires the world to be truly indeterminate! If God were to be present unveiled in creation then our independence and that of the whole universe would cease to be. A great deal of discretion must be applied in God's relationship with creation if the relationship intends the 'enabling of the beloved'. Chance, necessity, choice and relationships form the fundamental system by which freedom is bestowed on the cosmos by its creator. Such a partly mindless lottery, partly chosen course, is unacceptable to those looking for intelligent design. It is the mistake made by Darwin and many others who deny the existence of any God because of their assumption that the activities of a Creator will be detected by the presence in creation of purposeful design and progress towards identifiable goals by the shortest route. Instead of such a designer, Christians find God expressed in human terms responding and reacting to the other beings in the universe as a creative participant, humble and personal; dare we say, born in a manger. God, in Christ and through the work of the Holy Spirit, is seen as present as instigator, sustainer and participant, but not as designer, mechanic or puppeteer. Clearly we should not see Christ's action as an isolated event but as the process that focuses and brings to consummation a universal pattern of activity. This is a matter we will consider in the next chapter.

The consequence is that we see a cosmos in which God is aiding and abetting the development of ever deeper complexity leading to increased autonomy, up to and including human freedom and spiritual awakening in whatever form of creature and wherever that may have occurred in the cosmos. And here is the wonder, for the deeper the freedom, the more the cosmos is able to meet and commune with God and the more overt the relationship becomes. Most significantly, at the deepest level of living, the material and the spiritual come together; in the person of Jesus for example, or in the Church in the water of baptism and the bread and the wine of the Eucharist. The cosmos is the instrument of God's action and through it the Creator's intentions are made real. In the broadest sense the whole universe is imbued with God's grace and meaning, that is, it is sacramental, 'an outward sign of an inward and spiritual grace' as the old Anglican catechism puts it. The physical side of human life is no more and no less filled with grace and presenting grace than is the spiritual/mental/intellectual side. It is our task, as reflected most clearly in the eastern liturgy, to offer the whole of creation to its Creator in praise and worship. Our lives will find their true meaning and significance then as we express the enabling love that we have received, and live in proper relationships with God and creation as we are enabled to do.

Several points follow from this. Firstly, creation is of value to God in and of itself, not simply as a stage setting for humanity. Even if we (as creatures that are self-aware and can know and respond to God) had not evolved either here or elsewhere in the universe, creation still would have been loved and appreciated/valued by its Creator. Secondly, the universe has been endowed with the capacity and the freedom to become whatever chance, necessity and choice may determine. Thirdly, God is not managing the world in a hands-on fashion, manufacturing the specific products desired, but providing the freedom and opportunity for value to develop of its own accord, in a context provided and sustained by God. Fourthly, God is not an absent-minded clockmaker who wound up the clock and left it on the shelf to run all by itself. Rather, while providing loving freedom to 'be', God also takes part in creation as an agent, a participant in the drama of history, not as a puppeteer pulling the strings of history.

SPIRIT AND MATTER

If meaning is only to be found in what was meant, purposed or intended, then how are we to find meaning in our existence? If we accept that we are the loving product of the Creator through evolution, then the key to meaning for us must be based on God's intent in creation. We need then to listen to what creation has to say, and the insights of science are one of the most important sources that tell us what we are and where we are. We saw in Chapter 4 that we may discover ourselves as the sum of a context, relationships and a history.

The Christian faith we practise today is the consequence of many influences. This is no more clearly seen than in the attitude expressed by the Church and Christians at various times and places. From the Hebrew traditions we find an attitude towards creation that is quite earthy. God created the material universe, and it was good; God created men and women from the earth and they were also good. It follows that the material world and its expression in natural and proper ways are also good. Work and play, eating and drinking, sex and parenting are all good. Further, we are creatures of the earth; as evolved creatures, we are only truly ourselves in a material context and cannot be ourselves in any other! The material resurrection appearances of Jesus as reported by Luke ('He took [fish] and ate it in their presence', 24:42) and accepted by the Church, for example in the Nicene Creed ('We believe in …. the resurrection of the body…') recognise and confirm this reality. Such a reality, however, will be surely very different in many ways to the world of our present experience, not least in its freedom from the restrictions of time and place. Nevertheless whatever the future may hold for us must be expressed in a material form otherwise we cease to be human.

Another major source of ideas in the developing Christian community were derived from Greek thought of the first and second centuries. Here the material world was not considered 'good'. It was the source of all our problems. The real world was spiritual, and by some terrible mischance we, spiritual beings, had been trapped in flesh. The material world was the source of evil, pain, death and spiritual temptation. Further it was merely a shadow of truth which was only to be found in spiritual reality. To be saved we must be saved from the material world.

Such an attitude towards the material world leads to either of two outcomes, and while these were rejected by the Church in any extreme form, they are still present in the attitudes of Western society. Firstly, we may see ourselves as more than earthly creatures and aspire to the conquest of nature and ourselves. In the myth of progress that has sustained the dreams and aspirations of Western communities for the past two centuries we see a clear expression of this attitude. We are greater than nature, we are unstoppable in a conquest of all that is, including ourselves. Science, as a tool of scientism, has been a tool used in the furtherance of this dream. Materialism in its various forms has been one of its highest expressions. We see here the pride of our culture in all its glory. The tragic failure of this dream in our own time is manifest in our lack of self-control, in the making of yet more frightful weapons, of the sheer size of arms budgets, of our greed and rapacity towards the earth. There is no sign of nett progress in the human soul. Some things we do better than previous generations, some things worse.

The reaction to the failure of this dream has been profound. With our pride stripped from us we sulk and transform the failure of the dream of progress into the denial of meaning, the denial of purpose and the denial of any intrinsic value in humanity or nature. People, and the relationships in which they exist are commodities to be exploited and have only the meaning and value that we choose to give them. Economic rationalism and postmodernism express graphically the spirit of this age and can be seen, not so much as the beginnings of new and hopeful things, but as the ultimate working through into the human heart of the realities of materialism and the defeat of materialism, respectively.

Unfortunately, in deliberately confusing science with materialism in the mind of the community, those who support materialism have brought science into disrepute as part of the failure of the myth of progress.

The response we see to the failure of this dream might be called apathy. If we cannot conquer the world, if we cannot be more than mere material creatures, then we will be nothing. Let us eat, drink and be merry, let us not try, let us not bother as it does not matter; the fantasy worlds provided by heroin and the electronic chat room are as satisfactory places as any to live. The often desperate pessimism of such views is frequently combined, quite aberrantly, with an interest in superstition and magic.

Both pride and apathy are the result of our failure to accept the reality of our creaturely nature and the value of such a nature. We need to turn from the 'Greek' view of creation to the 'Hebraic' one. We are creatures; this was intended and therefore is right and good. This cosmos is where we belong. If our evolutionary origins and place in creation are taken seriously, along with the living relationship we can have with God and the clear value placed on us in Christ's actions, then we can avoid both the pride and the apathy.

To avoid these false images of ourselves we need to take seriously the basic fact of our evolution. Not only must we take it seriously but we must be glad of it. If we are to be redeemed from these misconceptions we must wholeheartedly affirm our creatureliness and our fundamental relatedness to the rest of nature. At the same time we must recognise that in us creation on this planet has taken another step forward into deeper complexity, and our search for meaning will succeed when we come to live consistently or in harmony with the purposes implicit in God's creation of the cosmos. If the existence of creatures like us is purposed as part of material creation then we must seek our meaning and purpose in the context of this extended creation. If in us creation has become self-aware of God then we will find our meaning in fulfilling our part in God's purpose in this extension of the complexity of creation. We will consider our place and role in creation further in Chapter 7.

6

REDEMPTION: FROM WHAT, TO WHAT?

REDEMPTION FROM WHAT?

While our world view may have changed in many ways due to the work of science, we still feel an often desperate need to make sense of our lives. For each individual this imperative takes a personal form. It may be the issue of life and death, of our personal identity and relationship to others, of the meaning and purpose of our existence, or again, of guilt and our moral standing before God or in our own eyes, that most concerns us. To varying extents we all feel the power of each of these issues, though their relative importance and urgency varies from person to person and, for an individual, from time to time. In some sense we all feel the need for 'salvation'; the need to save our lives from futility and fear. In the last few years we have come to recognise that science does not provide answers to these questions, though, by developing a more adequate natural theology, science can provide us with a far clearer and more focused material context in which to address these issues than was formerly available. We must also remember that previous theologies were developed in the context of beliefs and understanding of the material world held at the time of their development and the resulting errors need to be corrected. As Thomas Aquinas put it: 'A mistake about creation results in a mistake about God.'

A second dimension of this heartfelt need for 'salvation' comes from the real sense we have of being, at the same time, part of the continuity of the physical world and yet separate from it. In some way we feel there is a real discontinuity between us and the rest of nature. The evolutionary process that led from the inorganic to the organic and from thence to the human is a mixed blessing to its recipients. As the process continued, the original unity and simplicity of nature was torn apart, as we saw in Chapter 3, by the evolution of new levels of complexity with emergent properties, of attributes and potentialities not previously present (though, of course, implicitly possible). This increasing complexity has led, at least on this planet, to a creature that is self-aware, whose capacity to choose, to have freedom of action, is impossible in its less complex kin. Something new has come into the world. In us, nature has become conscious of its own existence and mortality.

It is a saying in zoology that embryological development of the individual reflects evolutionary development of the lineage. Our present dilemma can perhaps be seen in this light. As australopithecines were slowly transformed into humans there developed an increasing sense of self-awareness, a recognition of the real likelihood of personal death at some point in the future and a less instinctive link with — in fact a downright sense of separation from — the rest of nature. In a similar fashion, as the individual is thrust from the womb, makes its way amongst its siblings and age group, becomes more individualised at adolescence and ultimately approaches death, this sense of separation, of alienation, grows ever stronger. Life unfolds with increasing complexity and ambiguity.

We make ourselves through the decisions we take or that are made by others and affect us. As we grow, however, we come to realise that these decisions have moral consequences. We begin to develop a knowledge of 'right' and 'wrong' and the loss of animal, or childish, innocence. We experience estrangement from God, self, community and our wider environment. This slow passage from potential to actuality is, in the ultimate sense, the 'fall' so poetically described in Genesis, and it is a fall each of us makes from innocence into conscious freedom. It is 'original' in that in evolution our lineage made the same discovery that we each make as we grow. However, creation and fall coincide, as there is no point in time and space when the potential for choice between right and wrong was not available and would be expressed given a suitably complex creature. The freedom necessary for moral choice is implicit in the structure of the cosmos, though it may have taken fifteen billion years to become explicit in our corner of the universe. Unless one has the freedom to say no to God, one is not in the position to say yes either, and the freedom to say yes or no is an implicit part of the inheritance of everything in this universe. It is thus part of the loving intention of the Creator in creation.

How do we relate this sense of estrangement to sin, to those acts of disobedience to what we believe to be right? It is clear that we possess the capacities to deliberately act towards one another for good or for ill and that this capacity is not developed in a proton, a plant or a possum. Further, we can care for one another, we know we should care for one another and we experience remorse when we do not care as we believe we should. This capacity to label and feel that certain acts we have undertaken are either right or wrong is one of those emergent properties that set us apart from the rest of earth's creatures. Many higher mammals can certainly assess whether an action is advantageous or disadvantageous. In the apes we can see evidences even of working towards long-term social advantage. That dimension of human existence we call 'conscience', a sense of right and wrong, that reflects values we feel beyond those of our personal existence, even beyond this creation, is something truly human. In a similar fashion, many animals can be seen to work towards their own ends, but this is not what we mean by selfish behaviour, there is, we feel, a moral difference in our case.

In wrestling with the nature of our condition then, our theological reflection must take into account both our evolutionary identity as part of nature and our differentiation from it. We must also take account of the misuse we have made of our freedom. In practical terms and in our own time, these tensions can be seen in sharp focus in the ecological crisis our planet faces as a result of our behaviour.

We must first ask ourselves what we mean when we use words like 'redeemed' or 'saved'. By implication we often mean saved from something, saved from a situation in which we should not have found ourselves. Clearly there is a sense in normal usage that we would have been much better off if we had not needed to be saved in the first place, and our hope is to return to the state or place we were before. But is this always so? Often we are not the same people after a crisis that we were before it. Often our relationships are irrevocably changed as a consequence, and we do not really want to go back: we did not like the experience, but we would not want to give up what we have gained. Sometimes finding our way through a disastrous situation changes our relationships with others or our understanding of ourselves, and we would not like to see things return to the way they were. It is a frequent experience that our lives can be made deeper and richer in some ways because of such experiences. 'A place where it never rains is called a desert' according to the Muslim proverb.

If we see the need for redemption in such a light, as a process of being saved 'through' rather than just 'from' alienation, then we are not simply talking about salvation from sin or moral turpitude, for our search for redemption is a response to a far more fundamental situation, though redemption in the fullest sense can and must include a response to our personal and corporate misbehaviour. It must, more fundamentally, be related in some way to our place in creation, and our sense of ourselves and our mortality. Creation may be good, but one effect of our evolution has left us in a serious predicament. The very process of our arrival has led to a sense of alienation from the rest of nature. Alienation is not necessarily morally good or bad, it is simply a sense of separation or distancing from that to which we feel we belonged. In a constructive context it can lead to growth and a better integration, a more informed, less instinctive response to life. If not handled constructively however, it can be destructive, destroying what little peace we had. Alienation occurs whenever unity is broken; unity may be broken, however, so as to allow differentiation, as in birth, adolescence or death. Differentiation also allows other forms of response, as in exploitation of others or the environment, or self-centredness. Alienation is not equivalent to, though it includes the possibility of, sin.

Redemption seen is this light, then, is not a return to the innocence of our animal ancestors, nor is it the repression of the biological instincts of an advanced primate, for these have their proper place in the truly human condition as we saw the Hebrew tradition insist. Like other aspects of our total humanity, they can be misused and form the basis for sin, but they are not of themselves sinful. In a similar fashion neither the rational nor the subconscious dimensions of our minds are by their nature sinful but again, when misused or out of balance, lead to disaster. It is a challenge to our understanding and attempts to make sense of our condition that it is that which makes us human that can destroy us. Clearly we cannot go back to animal innocence, in that sense of being saved. We must go forward, dare we say, to become part of a new creation.

As freedom is implicit in the basic structure of this cosmos, so, of necessity, is moral evil, though both have only become explicit, as far as we know, in humanity. Consequently the need for redemption, seen explicitly in our need, is also present, though unspoken and without a voice, in the rest of nature. We use the term 'evil' also in another sense to describe the blind disasters and other chance calamities of nature, that is, natural evil. We can therefore see evil as wrongdoing or evil as suffering, and these lead us respectively to guilt or lament. However perceived, the opportunity and prospect of evil is a dark but essential side of the freedom needed for the emergence of new levels of complexity in nature, for the evolution of a new and deeper way for God and nature to commune. It seems clear that God cannot establish the context for the production of creatures that have real freedom of choice and simultaneously ensure that they will do exactly what they should do. The establishment of freedom risks, in fact effectively ensures, the presence of pain, death, cancer, earthquakes, fire and famine, as well as humanity's inhumanity. This may be the 'best' of all possible worlds that would allow freedom but, short of wishing for our own non-existence, we are left in guilt or lament to cry that the 'best' could have been a great deal better for those who have to live its reality!

If we truly find ourselves in a scenario in which a distant God has established a cosmos which includes freedom, but who is impervious to the pain of natural evil and of the moral evil that we commit, then we would be entitled to cry out in anger with the author of Job. Alternatively, we might give up in despair at the pointlessness of it all with Monod, the Nobel prize-winning biologist whose phrase 'chance and necessity' and the futility that he believed followed from it led him ultimately to suicide. This condition of God as an 'unmoved mover' stands as a moral challenge to any religious faith that sees God only in such terms. Christians, however, would claim that God is also a partner in this tragedy and that meaning flows from this relationship.

REDEMPTION BY WHOM?

There are, then, several points that arise from our discussion so far. Firstly, redemption is necessary for reasons that are far deeper than moral destitution, and that have been at least implicitly present since the big bang. Secondly, our fall is intimately related to what constitutes our humanity. We must also remember that we have the insights from science that the reality of this world is based not on things but on processes and relationships, and that death has been an essential part of biological evolution on this planet for billions of years. It is at this point that we strike difficulty with the traditional approaches of Western Christianity, at least since St Augustine's time, where great care is taken to place everything into categories, carefully defined, and it is assumed that our condition can be so defined. This approach permeates all of our thinking to the point where we find it difficult to conceive otherwise. He is a worker, she is a housewife. Even when we oppose the categories of thought we do so by proposing alternate categories.

It is interesting however that the early church, from Irenaeus on through the Eastern Church Fathers (that is, the patristic period, 2nd to 8th century) saw the matter differently. In their view, human nature is not seen as a static, unchanging category, that is, closed and definable. It is instead seen as an open, dynamic reality, determined at the most fundamental level by its relationship to its Creator, and at intermediate levels, open, by its development in a specific historical context in the world. They saw within this creative context, an unfolding in human lives and in humanity's existence that finds its highest point in the development of a fully expressed relationship with God. We might use the biological analogy of the genome of a fertilised egg which, over time, slowly comes to adulthood by the unfolding of its genetic potential in the context of the particular place in the world that it finds itself. The physical and psychological form created would be different if raised somewhere else (and in the framework of quantum physics described earlier, if it was possible to re-run the unfolding from exactly the same point in history, the result would be different again). This dynamism of the early, largely Eastern, church is in strong contrast to the static, categorical thinking that developed in the Western Church. This can be seen in the way such terms as 'nature' or 'grace' are used as categories in the West. These eastern concepts fit well with the insights of modern science.

In other ways the pattern of patristic theology is also similar to science. For example the early Fathers believed that there is Truth which we can live, but which we cannot easily define. In fact, we can more easily (and perhaps only) define what it is not. These theologians had a healthy respect for the unknowableness of God at the intellectual level (as opposed to the relationship level) and were content to leave open anything that was not seen as clearly wrong. Christianity was seen to be experienced both individually and in community, while Scripture and Christian tradition offered support and context. So also in science, the world is experienced, that is the reality, and we have a framework of interpretation inherited from those who have worked and thought before us, but this work and thought does not define reality, which remains a mystery. We also live with the consequences of laws, that we may never ultimately know, though we may know what they are not!

A patristic view of the world sees it as other than God, authentically itself. It however only finds its true self (or can only fully express itself, or find its meaning) in relationship to its Creator. Thus the world was not created in the form in which we find it, but is the result of fulfilling the God-given injunction 'Let the earth bring forth' (Genesis 1:24). Basil of Caesarea says, 'this short commandment immediately became a great reality and a creative word, putting forth in a way which transcends our understanding, the innumerable varieties of plants .... Thus the order of nature, having received its beginning from the first commandment, enters the period of following time, until it achieves the overall formation of the Universe.' The world therefore follows over time and in God-given freedom, its own process of evolution, development and growth and it is this world that science provides us with the capacity to study.

A further dimension of patristic thought from the earliest times is also relevant here. For the church Fathers, the fundamental issue facing humanity was seen as not that of sin but that of the reality of death. This is quite different from the view we have received in the churches of the West as intellectual children of St Augustine. In the East it is considered that only the free individual can commit sin. This is always seen as a personal choice and incurs the concomitant guilt. Sin is always a personal act not an uncontrollable act forced on us by our innate nature. If it was an 'act of nature' how could we be held responsible as individuals? Thus the sins of our forefathers (and mothers) is their responsibility, the concept of inherited guilt cannot possibly have a place in any just system, according to patristic thought. Of course the consequences of sinful acts can and do affect others because of the essential role that relationships play in being human. Similarly, the effects of the right acts of one individual affect others in the community, again through relationships. Death is seen as a universal disease that holds humanity in its power, materially and spiritually. The Eastern Fathers see sin as the consequence of the reality of death. In recognising our mortality we opened ourselves to inordinate attempts to satisfy the needs of mortal life. We break all rules for food, goods, pleasure, children, in an unavoidable and unsatisfiable drive to avoid death. The fundamental consequence of the fall in patristic terms then is death and its recognition rather than sinfulness. Our consistent drift into sinfulness is the consequence of our recognition of our mortality. This, of course, is compounded by the effects of the actions of others on our personalities.

Our consideration of evolution and the nature of the fall

however has led us to the recognition that death has been an essential element of life processes on this planet for at least three billion years. It is our knowledge of our mortality that drives sinfulness in the patristic mind and, if our earlier assessment is true, it is this knowledge that is a critical part of our fall from innocence.

For the early Eastern Church, then, it is our mortality, not inherited guilt nor an inherently sinful nature, that drives us to sinful action. We need to be delivered from the fear of death and, therefore, also from the necessity of struggling for existence. 'Do not be anxious about your life, what you shall eat or what you shall drink, nor about your body, what you shall put on. Is not life more than food and the body more than clothing?' (Matthew 6:25) The primary role of a redeemer then, is to destroy death, not to act as a substitute in a court of justice, though our subsequent behaviour needs to be recognised, admitted and dealt with.

What then does the early church have to say about our escape from this situation? From what has been said earlier about God 'out of time', and, in the previous section, regarding the nature of the fall in an evolutionary context, the work of Christ cannot be one of reinstatement of humankind as there was no former condition, except an animal condition, to be reinstated to! Clearly, the nature of our condition and the disastrous misuse of our freedom, were known and included in the original intent of creation (though in another, equally valid sense and from our own perspective, we had 'yet' to choose this path from those available to us). We see that this was clearly understood in the early church, for example Paul could write of Christ to the Colossians (1:15-18): 'He is the image of the unseen God, the first-born of all creation, for in him were created all things in heaven and on earth .... He exists before all things and in him all things hold together and he is the Head of the Body, that is, the Church. He is the Beginning, the first-born from the dead' or to the Philippians (2:6-7) '[Christ Jesus] being in the form of God, did not count equality with God something to be grasped. But he emptied himself taking the form of a slave, becoming as human beings are'. The way forward was seen by the early church as 'always' known in the love of the Creator, and there was no sense that we are looking at 'Plan 2' (redemption) because 'Plan 1' (creation) failed. Thus the coming of Christ in human form was seen to be both foreseen and foreordained independently of our tragic misuse of our freedom and was intended to deal with our recognition of our mortality and all that flowed from it.

Maximus the Confessor, who summarised much of patristic thinking, describes creation as a dynamic process intended to achieve the goal set for it in the loving kindness of God. This goal, that is, Christ, was both its initiator and its end. Christ was considered both its creator and, as its head, in the body of Christ (that is, the community of all believers), its final fulfilment. Through the process of Christ's becoming human and in obedience through death and so to the resurrection, we were seen to be given the opportunity to be made his own in the body of Christ. The incarnation, that is, Christ coming in truly human form as Jesus, therefore was viewed as an essential part of the destruction of death, for it could only be done within the freedom of creation, but also could only be done by God. The incarnation was seen as essential to salvation and the need independent of humankind's sinfulness and mortality. It was intended from the beginning as part of the process of creation. The obedience, death and resurrection of Jesus then, in whom both God and creation are expressed, was considered life-giving, precisely because both are present. We could not save ourselves and God could not negate our freedom by doing so by divine command without destroying the whole purpose of a creation based on freedom. It is God here as one of us that carries the day. Thus the cross and the empty tomb not so much provide a fit retribution for crimes against God and nature but vanquish the overwhelming reality of death, which has held us in its power, pushing us continually into a vicious cycle of sin and corruption. Just as the fall did not lead to inherited guilt, so redemption is not fundamentally about justification but of victory over death and futility.

Such a view of Christ's resurrection, then, means that the pervasive role death has in our psychological existence has, at least potentially, lost its hold on us. We are no longer slaves in the kingdom of death. Our future will no longer be decided by death. Life need not be pointless. Thus, we can move beyond the consequences of our fall as we accept the reality of, and become party to, the resurrection. Physical death of course remains in the present age and is still grievous, but it no longer, of necessity, dominates our thoughts and actions. Its significance and consequences are utterly different. As Meyendorff summarises the patristic position: 'Since death has ceased to be the only possible end of existence, humankind is free from fear, and sin, based on the instinct of self-preservation, is no longer unavoidable. The vicious cycle was broken on Easter Sunday and is broken each time the death of Christ is announced and his resurrection confessed.'

To change the language, we have seen repeatedly with increasing complexity in evolution, the appearance of emergent properties; a recurring pattern of new opportunities and unexpected laws. In the evolution of humankind we saw just such a jump to a new level, and this level opened to us, its recipients, the knowledge of our mortality which can be expressed in the language of a loss of innocence or as a fall into freedom. We have, in our recognition of our mortality, in our guilt and in our need for meaning, the capacity for a further step forward, and in Christ we find this, moving beyond the frame of the limitations of this material universe. We now have open to us the opportunity of moving beyond death and a time-limited existence. In Paul's words: '[The world] was made the victim of frustration, not by its own choice, but because of Him who made it so; yet always there was hope, because the universe itself is to be freed from the shackles of mortality and enter into the liberty and splendour of the children of God. Up to the present, we know, the whole created universe groans in all its parts as if in the pangs of childbirth.' (Romans 8:20-22).

Thus for Christians, Christ, emptied of power but not of love, became in time Jesus of Nazareth, the pioneer of their faith and the first of many brothers and sisters.

REDEMPTION TO WHAT?

A critical insight of science that is central to such a view of redemption is the key role that science gives to relationships in the processes of nature. From the breakdown of the separation between observer and observed in quantum physics, through recognition of the role of dynamic relationships in ecological systems, to the essence of being human, where our humanity is developed and expressed in our relationships, it is clear that relationships and processes have been recognised as central themes in creation as expressed at all levels of complexity. It would be expected therefore that, in the next step into the increasingly complex existence that is available to us through the very structure of creation, relationships would again be the key to the process but, as in previous breakthroughs to new levels of complexity, that relationships would take on new and unexpected dimensions. In the same fashion, as an extension of our experience in the natural sciences, we would expect to see in the emergent human condition unpredictable interactions, opportunities and dimensions as the way forward in God's intention.

Redemption, then, can be understood as about relationships and deepening levels of complexity, and this complexity will be far more than biological or even cultural. In Jesus, the opportunity for a far wider series of relationships with God is opened, and these begin the process of transforming us into the new creation (that is, the full expression of the next level of complexity, one attribute of which, for example, is that it is ultimately independent of time). In emptying himself and becoming totally human and, in his life, teaching and death, by acting in obedience to God, that is by maintaining an unmarred relationship with the Creator, Jesus can be seen as opening up the way into this new relationship. He becomes a transforming catalyst that can carry us into this new relationship in so far as we have a living relationship with him. In a sense, we can be seen as saved not by what he did but by what he is. It is in the making of the connection that we are saved. It is through our living relationship with Christ now that these new dimensions are opened to us. We have the opportunity to move into the fuller reality of the creation as conceived in the mind of God. In this deepened relationship we find and take our role in expressing the meaning for which this creation was intended.

Because of this different perspective, baptism can be, and is, seen by the Church Fathers in a different light from what we may be used to. While its role in the remission of sins is clear in adult baptism, that is, for those who have sinned through their own free personal choice, its fundamental nature is seen most clearly in the baptism of children. The Augustinian view of children being 'born sinful', not because they have sinned by their own choice but as an inheritance from Adam, is not accepted by the Eastern Fathers, in fact it is seen as nonsense. Thus, they see that the church baptises children not to 'remit' their as yet non-existent sins, but in order to give them new and immortal life, to transfer them into the living relationships of the body of Christ and the resurrection; that is, from the tyranny of death. It is transformation from the old creation to the new. It is the establishment of a new set of relationships in a new context and it is these changes that provide the opportunity for new and enlarged existence. Such sacramental actions, then, are not seen as miraculous breaking of the laws of nature but the ultimate manifestation, as far as we are concerned, of the ultimate Truth about the world and life, about humankind and nature, about the Truth that is Christ and life in Christ. Thus redemption is as much about relationships now, about life now as about life in some future place. It is the mechanism of transformation. In so far as it is a process, it is the process of transformation and provides us with a source for meaningful existence in the present chance and pain-filled world.

In the early Eastern Church, sin was best understood using the analogy of a disease, frightful and potentially mortal, rather than as a legal crime subject to sentence and punishment or forgiveness. For Irenaeus, God's reaction to sin is not anger at our rebellion but sorrow for our weakness. The central need is one of reconciliation. To say this is not to downgrade the seriousness of sin but to transform the nature of the images for describing it from those of the law court to those of the hospital. Thus, the recognition of the disastrous consequences of sin to the perpetrator and the victim, not to mention their relationships with God, is to open a yawning chasm of alienation in need of reconciling response. Such alienation must be healed by making atonement, by making 'at-one-ment'. Our part in this is the recognition of our behaviour and repentance. These are part of the process of reconciliation, of liberation and healing, found in the development or restoration of relationships within the body of Christ. Again the need to reject a vision of life in terms of formal categories shows through, with love, or grace, not seen as entities but as processes. Truly human lives are always seen as coexisting within the processes of material reality, for that is where we belong. Grace by itself does not exist, there are only grace-expressing people and events. To find grace we must look within the created world and its relationships. It is there we will find the carriers and occasions that are vehicles of God's grace and in these we will find the reconciliation that Christ offers.

Our real sense of incompleteness, of dissatisfaction, of fear and of guilt, may be taken as signs of the next level of complexity.

We know it is in our grasp, but not reached; that we have fallen short and in a thousand ways destroyed our own and others' capacity to move to the next stage. As Daly says: 'Jesus is God's self gift to creation, which in men and women has become capable of responding to God's offer. To be whole, or holy, is to be healed of the wounds inflicted by every kind of estrangement we experience in life: estrangements from ourselves, from our fellow human beings and from the environment in which we live. Estrangement from God occurs on one or all of these fronts and it is God who takes the initiative in offering and bringing about the healing and reconstructing process.'

7

OUR PLACE IN CREATION

We can now use these central, but largely theoretical, ideas and insights derived through theological reflection on the discoveries of science of the processes of the cosmos. How might they inform our decisions as we seek to live constructively in the world that science describes and we, at least in part, have made? It would be expected that understanding gained by combining the insights of evolutionary biology and Christian belief would be especially relevant to the critical and most urgent moral question facing humanity at the beginning of the 21st century: 'What is our relationship with, and our responsibility towards, nature (that is, the biological world) on this planet?' These are the issues we will explore in this chapter.

To many of us this claim may seem perhaps to overstate the situation, given the many other moral imperatives we face. The environmental crisis, however, is particularly acute for our generation, as wrong choices in the next few years will irrevocably destroy options for future generations. Several facts may help to bring into focus the immediacy of this crisis. Firstly, more than half the human beings that have lived since our species came into existence are alive today. Secondly, we now use 36 per cent of the total global land surface for agriculture. Thirdly, we are clearly using more of the available potentially renewable resources (for example, water, soil, forests) than we are replacing, to say nothing of the irreversible use of non-renewable resources. Finally, through the increasing loss of biological diversity in the forms of ecosystems, species and the genetic variation present within each species, we are wasting the heritage of nature.

These issues can be seen as of critical importance for the long-term future of humankind and, as our survival and welfare is of selfish concern to us, they are also of importance to each of us as individuals. However, they also raise serious moral issues that derive from our answers to the questions of who we are, what is nature and what is the relationship between these two entities. These questions have been present and relevant for as long as humankind has existed, but in our time they have been pushed to the forefront of concern by our behaviour.

Humankind, from the earliest times, has had a basically, though not exclusively, utilitarian view of nature, that is, nature provides the resources we need. Whether it is the management of native vegetation systems for the production of game and food plants through the use of a sophisticated pattern of burning the habitat (as used, for example by Aboriginal Australians), the slash and burn methods of early agriculturists or the high technology methods of modern agriculture, nature has been seen as providing resources for us to use. Parallel to this very pragmatic view, nature is sometimes personified and, as a consequence, appreciated for the gifts she bestows (including our lives), feared for her violence and capriciousness, and placated out of concern for the wrath that might come as a consequence of our behaviour towards her.

An alternate perspective has been the sense that nature simply provides a stage on which we live our lives. Just as a child may feel that parents and others are there simply to provide a context for them, we often have a similar attitude towards the world in which we live. The nature of this 'stage' varies from time to time. We may see it, quite unrealistically, as a lovely kindly place just made for us to live in, happy and comfortable. Alternatively, and equally unrealistically, we see it as a violent and degrading place, inimical to things of the spirit. For highly urbanised city dwellers, it has become very difficult to see nature as anything but a distant, vague, largely irrelevant, backdrop to busy lives. To view nature as nothing more than the stage setting for our lives has been a common attitude amongst Christians, who often have seen the world simply as the place where the supposed drama of the fall from perfection and its consequences, and the process of redemption from this fall, are worked through in human lives.

Another perspective, related to, but quite different from these directly utilitarian views of the cosmos, results from the aesthetic value nature has for us. We frequently enjoy being 'in' nature. We appreciate the beauty of a natural landscape or seascape and feel more alive when we find the time and space in our lives to relate to it. For many of us our annual holiday at the beach or in the mountains is an annual act of 're-creation'. In every culture, however, there have been some who have extended these ways of seeing nature, feeling them to be inadequate. Such people perceive nature in a different way, at the very least treating it as real and of value in a non-utilitarian way. In Western Christian culture such people as Eckhart and Hildegard of Bingen come to mind.

What material, then, can we draw on from previous chapters for theological reflection on our view of nature? Firstly, from the history of the universe we see that the universe is very large, has been here a very long time and is not a place in which humans can be seen as central. Its structure and mechanisms do not provide a place in which progress leading inevitably to humans on planet Earth can be seen as its purpose. Our reflections have led to the view that the cosmos is of value to its Creator, irrespective of the presence or absence of humankind. An alternate and commonly held view is that the cosmos has intrinsic value, that is, it is of value in and to itself. It is difficult, however, to ascribe intrinsic value to something, even something like nature, or the whole cosmos, as philosophically there must needs be a valuer for there to be a value. In a Christian view of the cosmos then, the valuer of nature is not us, nor nature itself, but its Creator. Nature cannot value itself, and its value is independent of our views of its worth. After all, we could hardly be the valuer, having been here for only 0.002 per cent of its history!

The process of evolution shows us that humankind is clearly derived from, and related to, nature. This conclusion immediately leads to the question: Are we only a part of nature, or something else as well? Clearly we are material beings; we cannot deny it. It is here that Christians again come up against their mixed Hebrew and Greek heritages. The denial of the value of the material world that we saw so permeating Greek thought, has left its mark on us: we are not comfortable with being material creatures, we search for what we believe are more valuable 'spiritual' qualities. We are not in balance. At the same time, science has led to the concept of emergent properties, and as we saw in Chapter 4, the spiritual dimensions of human life are emergent properties as they set us apart, in some sense, from the rest of nature. We are not just another species, but then neither are we independent of nature.

We also saw that we exist in and through our relationships. Our reality as creatures consists essentially of processes and depends on the quality of our relationships. Again we find a divergence between our Greek and Hebrew heritages. In Greek thought the word 'sin' meant to fall short of one's goal, to not fulfil one's potential. Sin was seen as that which stopped us reaching personal fulfilment or perfection. In Jewish thought sin was perceived differently. It signified the breaking of relationships, of disobedience leading to alienation. The latter insight fits well with the concept of ourselves expressed in relationships. The question then is how critical to our moral well being is our relationship to nature? If this relationship is damaged or incomplete, can we be fully 'ourselves'?

It is clear that morality is one of the 'emergent properties' that define the human condition. It is an attribute of being human that requires a personal response. Nature however does not have a moral sense. It is humankind, we who must make moral decisions. Nature just is. As with all emergent properties morality cannot be reduced to the laws of the next lowest level of organisation, in this case biology. We would not expect to find in biology therefore a basis for human moral behaviour, unless we assume humans are just another form of animal. For example, 'survival of the fittest individual or group' or 'leave more copies of your genes' cannot be defended as the basis of a moral code simply because it is part of the mechanism of evolution. We must look for such a code in the emergent properties of being human, particularly in the kinds of relationships that define the condition of being human.

The insight of science that we are derived from and have a basic relationship with nature, means that nature can be conceived, in a moral sense, as one of our 'neighbours'. Jesus clearly supported the Hebrew view of our responsibility towards our 'neighbour' as summarised in the command that we should 'love our neighbours as ourselves'. When asked who was our neighbour, he, through the parable of the Good Samaritan (Luke 10: 27-37), defined a neighbour as one who has compassion. Thus to live the moral law as Jesus understood it, we must become compassionate neighbours, in this we fulfil the need to live in true and full relationships. To do less is to sin in the Hebrew sense. Such sin alienates us, leading to hostility between creature and Creator, between creature and creation. In the latter case a proper response will require more than lip service to the rights and aspirations of other humans and of the rest of nature, given our continuing destructive behaviour summarised earlier. It will require a radical transformation in our thinking and major, one could say 'sacrificial', changes in our behaviour.

In a general sense then, what constitutes true compassion towards our neighbour, nature? We saw earlier that such love could be defined as 'enabling the beloved' and we examined this as the basis for the observed evolutionary process. We saw in Chapter 5 that the cosmos as the beloved of God was given the freedom and opportunity to explore its options and to become whatever it might be. Similarly then, we would expect that a loving relationship with nature on our part will be one that 'enables' nature, therefore we will need to consider what enabling nature would require. Further, we will need to decide whether our present behaviour towards nature constitutes compassionate behaviour and, if it does not (which would constitute sin in the full sense), what then we should do?

ENABLING NATURE

As our understanding of the evolutionary process shows, enabling nature would mean allowing it to continue to develop new forms and explore new options. Enabling nature, then, includes providing the physical space and resources that it needs to express its potential in this time and place. It seems clear to me that this does not mean humans using half of the world's resources. We will need to live more simply than at present, and there will need to be far, far fewer of us. This does not mean returning to the life of the 'noble' savage, we need to develop cultures and technologies that allow for fuller lives for a larger proportion of the human population, using far less of the available physical resources. We must move forward, and swiftly, in our understanding of nature and ourselves.

We also need to move forward culturally in our relationship

with nature. Unfortunately the 'tragedy of the commons' means that, if, as individuals, we change our ways then that portion of the community's common resources we refrain from using will be immediately gobbled up by those around us. As a consequence, though personal decisions by each of us about these matters are essential, the solution will require a fundamental spiritual shift in our perceptions as a community, of ourselves as part of nature, and the discovery that our fully expressed reality can only be in and with nature. The poorer we make nature by the simplification of ecosystems, the poorer will be our reality now, and the more serious the charge against us in the court of God in the future. We are in urgent need of a community world view; a 'myth' framework that articulates our contemporary understanding of the cosmos and also provides us with the spiritual framework needed at this time. That is, we need a myth that is 'true' for us, but more of this in Chapter 8.

INTERGENERATIONAL EQUITY

An interesting feature of courses in environmental science in universities these days is that they include an explicit consideration of ethical issues. While an ethical framework is implicit in the methodology of science, for example that data presented were really collected and are fairly and honestly presented, it is not felt necessary to make this explicit in university degree programs. The situation in environmental biology, however, is different, for ethical issues are considered to be the basis for action in our relationship with nature, rather than simply an essential part of our methodology. A central theme of environmental ethics can be summed up in the phrase, 'intergenerational equity'; that is, we should not use resources in such a way as to restrict the options of future generations. As the World Conservation Union (IUCN) puts it, in a thoroughly utilitarian way, in the World Conservation Strategy: 'The management of human use of the biosphere so that it may yield the greatest sustainable benefit to present generations while maintaining its potential to meet the needs and aspirations of future generations'. In short, we should leave the world intact and in good order to our children and grandchildren. It is apparent again that this can be seen as another extension of the moral injunction to love our neighbour. Here our neighbours are our human neighbours in time, and our enabling action is to leave them a world that is not even partially destroyed. Again, to break faith in this relationship is sin, as defined above, in the deepest sense.

In the two contexts given above, then, we are faced with moral (as well as utilitarian) imperatives to act in such a way as to enable nature to thrive and to express itself to the fullest possible extent. It is clear from all the evidence scientists have collected and made available to the community in the past fifty years, that our behaviour towards nature is truly immoral and we can expect to have to answer for this, not only to our grandchildren but in the court of God. To see our behaviour as a community and as individuals as no more than a technical problem of overpopulation and overuse, is to miss the point and to make a real solution psychologically and spiritually impossible. The space needed to enable nature to continue its full existence will require more than minor changes in our behaviour; it will require a radical transformation of our attitudes, and sacrificial changes in our behaviour. We will need to make reparation for admitted past misdeeds. In short we will need to undergo that traditional Christian process called 'repentance', which is based on the key concept of 'changing direction' (of our lives). Such a process means firstly admitting the reality of our behaviour to ourselves, then admitting this to God, that is taking personal responsibility for our actions, and finally establishing new goals and patterns and acting on them. To do this we again will need a 'true myth' to live by. We will need to go through this process both as individuals and as communities. Only then will the changed relationship with nature allow us to live truly human lives, that is, to live our lives in their true context.

OUR ROLE AS PART OF NATURE

If we accept that we are a product of a long process of evolution and are closely related to all the species that share this planet with us, and if we also accept that we are most fully ourselves in relationships, then it would be not surprising if we feel spiritually at home in some distinctive way when we find a quiet spot in the mountains or by the sea. There is a 'rightness' in this, a realisation of the connectedness of our common existence. We find such places beautiful without the need to anthropomorphise nature or make it into a god. Such places just 'are', and sometimes, for a little while when we visit them, we just 'are' too. We can be glad in creation; there is a timeless sense of belonging, and we can worship God, really without words, just 'heart to heart'. We can simply enjoy our existence and the existence of the world around us. This world however is mute; it can do no more than 'be' before its Creator, and here we find a special place in nature for humankind. The special attributes of our species allow us to speak the words of praise and knowingly make the offering of all creation to its Creator on behalf of all creation.

The church in earlier times knew this and so we find, for example, that this role is a key feature of patristic liturgy. Eastern Christians saw our worship as being on behalf of more than ourselves, for we can comprehend and speak out of our insights, about our, and the world's, condition. We can, however, speak truly only if we take our place fully in the relationships of the world, not if we try to dominate it or see ourselves as fundamentally independent of it. We may have a priestly role to speak for, to represent, all creation but we can do this only as part of it.

We also have the right and privilege to enjoy nature not only as creatures in creation, but in our understanding. Scientists, irrespective of their beliefs, are mostly driven on by their curiosity, their desire to know and to understand. There is an indescribable excitement, an exhilaration, a wonder, when our work suddenly shows us some new insight into the structure of nature. We stand in awe of the world, in what I believe for most would be a form of true worship, for, as Meister Eckhart puts it: 'Every creature is a word of God and a book about God'.

The consequences of lack of recognition of our place in nature found generally in modern Western culture can be seen in the attraction many European Australians feel for Aboriginal Australian spirituality, and in parallel behaviour elsewhere in the world. While it is not possible for most European Australians to take to their hearts the mythology of the belief system of Aboriginal Australians, for example their mythic connections to particular aspects of the landscape, the felt need is real and reflects the abject poverty of the present spiritual position of Western society with regard to the cosmos. The arrogance of a claim to be utterly other than nature, to consciously or subconsciously deny our fundamental relationship with nature, underlies our spiritual dis-ease. We need to discover a solidarity with nature by finally accepting that we are part of it, as Aboriginal culture does, though we will need to do it in our own way in our own science-based and largely urban culture. I am sure that we will need to find physical, emotional and intellectual ways to make this insight and commitment real in our daily lives and to express it in our worship. We will also need to find ways to do this before we can truly begin to change the behaviour of our culture towards nature, given the material changes and costs that will be the consequence of such changed behaviour.

As we discover and joyfully accept our creatureliness and our co-dependence with nature on God, we find that we have a special bond with the other creatures of the world. If we do not deny our inheritance but accept it with joy, we can give up our pretensions to be more than we are. It is not our world, in the sense that we are not its owners. It is, however, a world of which we are part. As we reflect on and understand God's purposes we can validly take up our role in the creative work of God in the world. We are not, after all, just passive observers of the cosmos, we have creative roles to fill!

We find, then, that in us nature is able to express itself towards its Creator in ways that it could not before. This does not mean of course that there was no relationship between cosmos and Creator before humankind; it was present from the first moment. It is now however a far more meaning-filled relationship, as creation in us can now accept God's invitation to deeper intimacy. Through evolution, the cosmos has always been a co-worker and co-creator with God, creating and destroying wonderful faunas and floras over and over again. For example, when the world of the dinosaurs was almost completely destroyed 65 million years ago, new faunas and floras evolved within only a few million years, so beginning the world of mammals we know today. In us, however, this work of creation and destruction can and does happen consciously and we can intentionally create or destroy. The freedom we have allows us the opportunity to refuse as well as to accept the invitation of love; the invitation to enable. We can also accept or refuse God's loving and enabling assistance provided through Christ's work and the sustaining presence of the Holy Spirit.

OUR ROLE AS OTHER THAN NATURE

As well as the need to see ourselves as a part of nature, we have also to recognise, as we saw in Chapter 6, that the level of complexity and emergent properties, which makes us what we are, set us apart from nature in distinct ways. We now need to look at what these differences mean for our behaviour towards nature.

One of the key attributes of love can be seen when the beloved is other than self. I am me, you are you, and we are in love with one another. To love, to enable the beloved, is not to make them what we are, or even want to do so, but to offer those we love the opportunity to become what they can be. Individuality is increased in an effective loving relationship, not decreased. In our relationship with nature then we would expect to see its individuality increase.

As we saw in Chapter 5, we have found that fundamental attributes of the cosmos and its creatures are to be independent, to explore and to choose. We would expect, consequently, that fundamental to our enabling actions towards nature would be to provide the opportunities for increases in the diversity of landscapes, in the number and range of species, and in the interactions between the species that share our home world. We would expect to see a pattern of humankind working with nature, neither ignoring or dominating it. However, we do not see such a pattern, we see instead a steady reduction in diversity and opportunity for nature. We choose, but too often we choose destruction or homogenisation. As a consequence of our actions we see an increasing rate of simplification of ecosystems and decreased differentiation across the planet. We should instead be seeking to understand and support nature better and we should be glad to see the creatures of this world exploring their reality.

A sense of reverence can and should be seen as a central value in the relationship between humankind and nature. In this situation reverence results from a deep and due respect for our relationship with nature. This reverence is seen in the behaviour of scientists in their work. In the face of creation in all its complexity, simplicity and elegance, scientists feel a sense of awe, that is, excitement, amazement and wonder in what they have discovered and this reverence adds value to their lives. This awe and recognition of indissoluble relationships in the cosmos can form the basis for a morality, or perhaps an essential ingredient of a religion, that truly values the cosmos. We feed our imagination, our very souls, on the sacredness of the world.

Further, if we changed our perspective, we could enjoy the world more nearly as God does, especially its complexity and capacity to explore possibilities. As a scientist I feel I have a particular role to play in this and a special appreciation, both at the intellectual and aesthetic levels of appreciation, of the created and evolving order. As a consequence, I feel the urge to encourage and nurture the complexity and diversity of the world. All scientists also have the responsibility to make this beauty and wonder known to the human community that supports us and our work. It is the least we can do!

It is an easy but fatal step to see our special gifts as allowing humankind to claim domination over creation. However, what we have is a derivative power only. As created beings, we do not own the world or anything in it. If we wish to act for its Creator towards creation then it can only be as co-workers with God. Perhaps we might see ourselves as 'stewards' using our skills on the 'owner's' behalf.

In an ancient household, the steward was both a member of the community and set apart from the other servants. This difference was clear in that, unlike other members of the household, the steward was directly responsible to the owner for the decisions taken. Stewards knew and understood the situation related to their responsibilities and acted according to their best understanding. On the other hand, the steward was a member of the household and there were no absolute rights attached to the position. Our responsibility, then, if we see ourselves as stewards of the Creator in nature, is towards the biological community of which we are a part, and should take the form of further enabling that community to fulfil its role as intended by its Creator.

Our context, however, is more than the cosmos, for our context is as creatures knowingly living and working before and with their Creator. It is here that the sustaining and sanctifying work of the Holy Spirit can provide insight and assistance. As we work with our Creator and the cosmos we find the Holy Spirit, she who was called 'Wisdom' by the Hebrews (for example, Proverbs 8), acts to enlighten our choices. If we recognise and respond to these relationships, we will recognise that attempts to dominate or discount nature are not appropriate stances for creatures of God to assume and we can begin the process of identifying what appropriate stances might be.

Similarly, our differentiation also adds further dimensions to our roles as priests of creation, that is, as well as offering creation to its Creator, we can present God to creation. Again we can see that this should be an enabling action and we need to work through what such an action and offering would be in practice.

In my view we are far from having any real idea what these roles would entail, nor do we have more than the beginnings of the skills needed to carry out our tasks responsibly.

The first problem we face with such ideas is that they are couched in the language and imagery of another time and place. They do not strike to the heart of the average Westerner. Where, however, are we to find a source of adequate or 'true' words and images that will bring these realities to life for us? Images that will provide us with the necessary spiritual commitment to carry through to completion the needs of love for our cosmos? Neither science nor religion alone have the vocabulary, but together, perhaps they might. The almost systematic practice of keeping them intellectually and linguistically separate has made it very difficult to begin such a serious dialogue. Christianity has also failed to reclothe the spiritual experiences on which it is based with the insights, language and imagery of our culture; rather it continues to flee to the worlds of 1st century Greek culture or 12th century medieval culture.

At the practical level, many of the problems we face are known and people are at work attempting to find solutions to them. Trying to identify the right way to apply the possible solutions, however, requires a proper ethical and spiritual perspective.

The problem of the deep malaise we face at the ethical and spiritual level needs deeper consideration by our community. Firstly we will need to be brought to an acceptance of the reality of our misbehaviour. Only then will finding a suitable cultural basis for living be possible and this will be a work that we will need to address through finding community answers to the questions of who we are and of our relationship with nature. As we seek to replace the 'myth of progress', I believe that such answers will require us to find and accept a fundamentally different spiritual framework, one that will provide values, meaning and purpose to our lives in an evolving world and the guidance needed in the development and use of appropriate lifestyles and technologies.

Anachronistic as the terms may be, it is time we took seriously the moral and spiritual dimensions of our place and roles as stewards of the earth and priests of creation so as to enable the world in which we live.

8

CREATION

Myths tell the sacred history of creation or of the beginnings of things. They describe the events that occurred at the time that all, or specific aspects, of reality came into existence. Because they are meant to be history, myths must be 'true'. That is, in so far as they deal with and explain observable realities, processes and historical events, the statements they make about the material world must be accurate. In some sense then, Chapter 3 could be described as, or at least used as, a myth, as it describes the history of our creation. Attempts have been made to transform such science into myths to live by. As a myth, however, evolutionary history by itself is inadequate because a myth must include other dimensions and attributes. For example, by knowing, or more accurately, living the myth, one can enter into the rhythm of reality as it is/was created. In a real sense, one is seized by the powers involved in creation and takes part in, and is part of, the original events, as well as living in the outworking of those events in everyday life. When this is done, the world and its activities take on a spiritual reality beyond those of simple material existence. As a result, the participant/believer is provided with a meaningful context in which to make sense of the world and a paradigm for decision making in an otherwise ambiguous reality.

We in the West have been nurtured for generations by the two great myths we have inherited; that in the set of Hebrew stories found in Genesis and that of 'progress', that is, of humans mastering knowledge, the world and themselves as gods in their own right. Unfortunately these myths have lost their life-giving power as we have come to recognise that they are not 'true'. They do not fit all of the 'facts' as we have experienced or discovered them.

Religion and science, however, together provide images and truths that do need to be integrated and expressed in any valid system of beliefs for our time. Below I have included an example of my own struggle to find words that reflect the truth that can be lived. They are clearly inadequate; I hope they may challenge someone with more insight and ability to show the truth in a mythic form that will catch the heart of our community and give us a better basis for developing our relationships with creation. As always truth is a work in progress.

A STORY OF CREATION

'I have no children,' God said again and again, 'I have no children. We need to love and cherish and, truth to tell, we need to be loved and cherished in return.'

'We are yours to command!' God's messengers replied.

'We need partners, those who are not us, those with whom we can share and love and dance' said God.

'It cannot be us then' they replied, 'for if they are to be distinct, then they must be independent'.

'Yes', God said, 'this kind of love can only be shared with those who are able to freely listen, or not, and choose to respond, or not.'

'But you are so all-encompassing, how can such creatures be?' some in the court of heaven asked.

'We will need to create a separate place, where they can become themselves. A place that will explore its own reality and make of it what it can, where blind chance and free choice are possible.' God replied.

'Chance! Choice!' exclaimed the court in shocked tones, 'but that means they may choose to be other!'

'Yes that follows' God replied. 'This will be very costly and you will be hurt' they warned. 'Yes' God said quietly, 'I know, but we will love this place and all the children that grow in it.

Let it begin!'

In the beginning time awoke, and stretched, mass also woke and looked about.

Seeing they were now awake, God's Wisdom came to lead them into the dance.

She came and taking them by the hand showed them the first steps they should take.

They moved forward and found the steps were right and natural and good.

As they danced together the steps became more complex and each step took on a life of its own.

They were glad and sang their joy in the music of velocity and gravity,

and were confident in the stability of universal constants.

As they watched together they saw that each step, no matter how small, became the beginning of its own, increasingly complex, dance.

As they danced, Wisdom danced with them, and She led them into new and more complex patterns.

Suddenly they realised that Wisdom was not alone but She too had a partner, and He was beautiful and filled with joy and it was from Him that they had come.

From the moment of creation the cosmos was filled with chance possibilities. In the first few moments processes occurred that determined the nature of the Universe for its entire history and that made it irrevocably different from any other possible universe. By the ordering of these early events, and in combination with the values set for the universal constants, the possibility of galaxies and stars derived. Chance differences in density led to such galaxies and stars. These followed paths of their own and there were collisions and the destruction of some of the patterns formed. In time these first stars ran low on fuel and some collapsed and exploded, spreading heavier atoms across space. New stars arose, and as the cycle of stars was repeated, the levels of heavier atoms gradually increased. Eventually, after seven billion years, in no special region of a galaxy, one particular star was born, and its planets. One of these planets was made from the heavy elements forged in the first stars and, as it happened, it was at a suitable distance from the sun for water to neither freeze nor boil across most of its surface. Over another half a billion years the air and water and land of this planet produced molecules in more and more complex forms. These came together in systems and in time, these systems could be considered alive.

'Did you see that!' God's messenger cried. 'Yes' another replied, 'it chose'.

But then the small creature was accidentally destroyed, and the court of heaven sighed.

'There will be others that can choose' Wisdom reassured them as She danced on.

And there were, and the dances of these were more complex than anything seen before, and became ever more so.

After several billion years, the world's oceans were filled with life and beauty as well as pain, death and decay. For increasing complexity led to increasingly sophisticated nervous systems, and neural reflex reactions to escape harm gave way to complex behavioural responses, and pain became real. As each individual and species faced the chances of life and made its choices, new opportunities appeared. The lush green terrestrial world was still uninhabited but some creatures found ways to live there and take advantage of the opportunities it offered. With increasing complexity in certain lineages came fear and pleasure, social structures, planning and the use of tools.

'Is this world not beautiful and good?' God asked. 'But there is pain and hurt here!' God's messengers objected. 'How can it be worth it?'

'Look and you will see, for how can freedom exist otherwise?' God replied.

And they watched and waited and pondered the history of another half-billion years.

They sat quietly drowsing in the sun, as Wisdom came and sat with these apes that were to be no longer apes.

'Welcome' She said and as She kissed them, they awoke and looked about and saw the world with new eyes.

'I did not realise' the woman said, looking at the carcass they had been eating from, 'that this was alive and now it is dead; as are my mother and your father. It will come to us too!'

In horror and fear the man and the woman huddled together and he said. 'You are right, we must leave this place and find a place that is different, where there is no death and we can live forever.'

Wisdom rose to speak with them as to what must be, but they turned away, eyes blinded by fear, not seeing her nor the world from which they had sprung, nor even truly seeing each other. They fled on and on, always looking for somewhere better and arguing as to where to go and what to do.

Wisdom called to them and every now and then one or other

of them would hesitate and half turn and listen to what She was saying.

While continuing the dance of creation, She persisted in calling to them, and to their children, and to those that came after, for as long as their world exists.

And sometimes they remembered some of Her words, but more often they were angry or frightened, and forgot, and struck out at God, the world and themselves. Nevertheless those who were less afraid and angry, did hear a little of what She had to say about their Creator, and love, and relationships.

'They are destroying themselves and each other, and the world of which they are a part and for which You have poured out so much love!' the court of heaven cried angrily, turning to God. *'Why do you not stop them?'*

'It is their right not to listen to Wisdom,' the Creator replied, *'it is their right to fail.'*

'There are however some who do listen and we shall invite them to try the next step in the dance.' God declared. *'We will show them the true nature of Service. Those who can, will join His dance and discover what He and they can become so that we can care for each other.'*

GLOSSARY

To aid the reader, brief information on some technical terms and individuals mentioned in the text are given here. Several words with more than one meaning are also defined.

adaptation A heritable aspect of an individual's phenotype that improves the chance of the bearer surviving or reproducing in the studied environment.
allele One of the forms of a gene.
amino acids The twenty different chemical building blocks used in proteins.
Augustine of Hippo (AD 354-430) Renowned for his massive intellect and his penetrating insight into and understanding of Christian experience and belief. His influence on subsequent theological development in the Western Church, whether Catholic or Protestant would be hard to overestimate. His explanation as to why God is outside time is given in *The City of God* Book XI, Chapter 6.
Basil of Caesarea (AD 330-379) A holy, learned and statesmanlike man who was an important participant in the development of traditional Christian belief in the period leading up to the production of the Nicene Creed.

chaos The behaviour of physical systems where the effects of an input are not simply related to the output. Very small changes in input may have very large effects on the way the system functions, or, conversely, large changes in inputs may make little difference to function. The effects are so subtle as to be truly unpredictable.

chromosome A linear end-to-end arrangement of genes in a package.

Community of faith The fellowship of all baptised Christians in all time and places.

constant, universal A number expressing some property of the universe that remains the same under all circumstances, for example the gravitational constant.

copying errors Mistakes made when making copies of genes for daughter cells.

Creed, Nicene The Church Council held at Nicaea in AD 325 developed and authorised a concise and formal statement of the profession of faith which is widely considered a general measure of orthodox Christian belief in both the Eastern and Western Churches.

dating, radiocarbon The estimation of the age of an item from the proportion of carbon-14 remaining in a biological sample and the rate of decay of this isotope. The method can be used for estimating the age of an item containing carbon up to 40 000 years. For longer time periods radioactive isotopes of other elements with slower rates of decay must be used.

DNA A double strand of complementary nucleotide bases. The order of the nucleotide bases provides the coded information needed for making proteins and other gene products.

DNA sequence The order of the bases in a piece of DNA.

Down's syndrome A human genetic condition caused by the bearer having three copies of chromosome 21 in each cell instead of the usual two.

Eastern liturgy The form of the Eucharist, Mass or Holy Communion services used in the eastern orthodox churches.

El Niño southern oscillation (ENSO) Intermittent warming of the waters of the Pacific Ocean off Peru. This is related to radical effects on the weather through changes in the pattern of movement of water and air masses in many parts of the world.

embryology The study of animal and human development from the fertilisation of the egg to birth.

emergent properties Laws and patterns seen in complex systems not seen in the simpler systems from which the complex system is derived. These properties cannot be predicted from a knowledge of the simpler parts. For example, the laws of genetics cannot be predicted from a knowledge of the biochemistry of DNA.

entropy The degree of disorder of a system. The total entropy of any isolated system cannot decrease in any change. The degree of entropy in one part of a system can be decreased only by increasing the entropy in other parts of the same system.

eucaryote The more complex forms of cells found in plants and animals compared to the forms found in bacteria.
Eucharist The supreme act of Christian thanksgiving. Also called the Holy Communion, or the Mass.
faith Reliance, trust or belief such as affects character and conduct and is not amenable to material proofs.
Eastern Church fathers A group of ecclesiastical authors up to the 4th century whose authority carries special weight due to their widely recognised orthodoxy of doctrine, holiness of life and the approval of the church.
gene The fundamental element of inheritance. Usually it can be equated with the instructions for producing a particular form of protein (including when and where).
genetics The study of genes and mechanisms of inheritance.
genus A group of species of organisms having a suite of characteristics in common derived from a common ancestor.
God Used in this book in the normative Christian sense: The creator, redeemer and sustainer of the universe and the object of supreme adoration.
grace Unmerited assistance provided by God that leads to sanctification, virtue or excellence.
haemoglobin The red protein found in blood that carries oxygen from the lungs to the rest of the body.
hard selection The form of selection where the mutant form of the gene directly affects the fitness of the genotype to such an extent that the individual carrying it leaves no offspring.
Hildegard of Bingen (1098-1170). A Benedictine abbess famous as a visionary, artist, musician and observer of nature.
Hominidae The taxonomic group (family) to which all species of the genus *Homo* (including modern humans) belong.
Holy Spirit The active essence of God and the instrument of God's creative, inspiring and sustaining action in the cosmos from the moment of creation onwards. Personified as female and conveying wisdom and religious understanding in later Jewish writings.
hypothesis Proposal put forward to explain observed facts.
immunoglobulins The proteins involved in the production of antibodies against infections in humans and higher animals.
incarnation The process whereby the eternal Christ takes on a fully human life as Jesus of Nazareth, thereby being both truly human and truly God.
inclusive fitness The relative probability of a genotype leaving offspring either via its own offspring or by offspring of related individuals.
induction The process of inferring a general principle from a series of examples or cases.
intrinsic value Of value in, of and to itself. This value is not dependent on any external agency or comparison.

intuition The subconscious process of obtaining a direct insight into the nature of a process without the use of reason.

Irenaeus of Lyons (c.130-200) He forms a critical intellectual link between the early church and the catholic and orthodox churches that were derived from it.

Jesus of Nazareth (c.6 BC-c.AD 28) The central figure of distinctly Christian belief, where he is held to express in one person both God and human natures and who by his obedience in life and death and through his resurrection, offers deliverance from death and sin to those who will accept it.

law A well established principle established by observation and known to hold over a wide range of conditions.

liturgy Here, the form and structure of the Eucharist, Mass or Holy Communion services.

Macaques Monkeys of the genus *Macacus* that have been studied intensively because of the development and evolution of cultures within and across their social groups.

materialism The belief system that nothing exists except matter, its movement and modification and that consciousness is wholly due to material causes.

Maximus the Confessor (AD 580-662) A Greek theologian who summarises much of Byzantine Christian thought.

Meister Eckhart (1260-1327). German mystic and monk. One of the most famous preachers of his time.

microevolution Evolutionary processes occurring within species.

mitochondrion A part of all eucaryotic cells. They are distinctive in that each has its own genetic system. Believed to be derived from bacterial cells that parasitised more complex cells a billion years ago.

modernism The faith that true knowledge is certain, objective and good, and accessible to the human mind. Reality can therefore be understood using reason and our insights into reality can be assessed using it. Modernism assumes that it is possible, under certain circumstances, to obtain objective knowledge about reality. It is usually optimistic in outlook and associated with belief in progress in human understanding, welfare and happiness.

monomer A molecule that can combine in a linear fashion with many copies of the same or similar molecules to form a polymer.

morphological characteristics Attributes of the physical structure of organisms

mutagens Agents that cause increased rates of change in the genetic structure of genes, for example, many chemicals or radiation.

mutations Altered DNA sequences in genes.

mystery A truth not open to analysis by the use of reason.

New Testament The set of books found in the Bible and prepared and accepted by the early Christian church as central to the Christian faith.

Noahian Flood The worldwide flood described in the story of Noah found in the book of Genesis in the Bible.
nucleic acids The basic building blocks of DNA and RNA.
paradigm, dominant The pattern of scientific belief/assumptions held by the majority of scientists working in a discipline at a certain time.
Patristic Period The time in Christian history between the close of the 1st century and the 8th century.
phenotypes The forms taken by some characteristic of organisms and reflecting variation in the underlying genotypes that control their development.
polymerisation The process of making a polymer from monomers.
polynucleotides DNA or RNA.
postmodernism This is not an organised belief system but a diverse and conflicting set of ideas that have in common only that they question the assumptions of modernism. Proponents believe (if such a word can be used of what is basically a highly pessimistic non-system of unbelief) that truth is not certain, there can be no such thing as objective perception, nor is there such a thing as good. The rational mind is not considered the only path to truth and even when found, truth is relative, indeterminate and participatory, that is, 'true' only to the community that finds/accepts it.
priest A mediator between God and the world (acting in both directions).
primates The lemurs, monkeys, apes, humans and their relatives.
quantum physics The area of physics derived from the wave mechanics description of the nature of particles and the Heisenberg uncertainty principle (that is to say, it is impossible to determine both the position and momentum of a particle).
reason To reach conclusions by the application of logic to premises in a consistent way.
reductionism The belief that a system can be fully explained in terms of its constituent parts.
replication The production of a copy of the DNA within a cell.
resurrection The belief that Christ was raised from the dead to a new form of life three days after his crucifixion and death.
RNA A single stranded nucleic acid related to DNA. Can be used to carry the messages coded in DNA to other parts of the cell for use in the construction of proteins.
sacrificial Behaviour damaging to personal interest for the sake of some desired object.
scripture The sacred writings of the Old and New Testaments, the Christian Bible.
selection Changes in allele frequencies between generations due to differences in fitness between alleles.
sin Behaviour leading to the breaking of relationships; disobedience to moral imperatives leading to alienation.

soft selection The pattern of selection where the mutant form of a gene directly affects the fitness of the genotype only when the bearer is in competition with organisms bearing alternate forms of the gene. In their absence individuals carrying the mutant form survive and breed quite satisfactorily.

species Populations of actually or potentially interbreeding individuals that are separated by reproductive isolation from all other populations.

speciation The process by which a single species is divided into two or more new species.

theology The rational study of religious faith.

theory A system of ideas based on general principles and consistent with all relevant data.

Thermodynamics, Second Law of See entropy.

Trinity In Christian belief the union of three Persons (Father, Son and Holy Spirit, or Creator, Redeemer and Sustainer) in one Godhead.

uniformitarianism The belief that geological and other physical processes are the result of forces acting in a regular and continuous manner.

utilitarian value Value measured in terms of usefulness or serviceability.

value The worth of something relative to other things or to some common medium of exchange.

Y chromosome A chromosome found in male mammals that is inherited from fathers to sons and contains the gene that instructs a foetus to develop as a male rather than a female.

FURTHER READING AND BIBLIOGRAPHY

The purpose of this list is to provide sources for those wanting to follow up particular issues. I have attempted to avoid the specialist literature and selected material accessible to the ordinary reader. In some cases I have added comments as to why a book has been included. I have tried to include a range of references that make markedly different interpretations of the basic facts from those presented in this book (and with which I cordially disagree). The material is given by chapter and the sources of the various quotations, or the ideas attributed to particular authors are unambiguously given. I have avoided books that deal with or use information in my areas of competence that have made basic errors of fact. Books have been listed only once, though they may be relevant to several chapters.

PREFACE

Huxley, TH (1931) Autobiography. In *Lectures and Essays,* Watts & Co, London, pp vii-xix.

1 THINKING ABOUT THE WORLD: SCIENCE AS A METHOD

Charlesworth, M (1982) *Science, Non-science and Pseudoscience,* Deakin University Press, Melbourne.

Charlesworth, M *et al.* (1989) *Life amongst the Scientists: An anthropological*

study of an Australian scientific community, Oxford University Press, Melbourne.
Gould, SJ (1999) *Rock of Ages: Science and religion in the fullness of life*, Ballantine, New York.
Grenz, SJ (1996) *A Primer on Postmodernism*, WB Eerdmans, Grand Rapids, Michigan. A Christian introduction to the main ideas of postmodernism.
Grinnell, F (1992) *The Scientific Attitude*, Guilford, New York.
Medewar, PB (1972) *Induction and Intuition in Scientific Thought*, Methuen, London.
Sober, E (1993) *Philosophy of Biology*, Oxford University Press, Oxford.
Sokal, A & Bricmont, J (1998) *Intellectual Impostures*, Profile Books, Hatton Garden, London. Summarises some aspects of the misuse of science concepts by postmodernists.
Winsor, F (1958) *The Space Child's Mother Goose*, Simon and Schuster, New York.

2 THINKING ABOUT GOD: THEOLOGY AS A METHOD

Macquarie, J (1966) *Principles of Christian Theology*, SCM Press, London.
Torrance TF (1985) *Reality and Scientific Theology*, Edinburgh, Scottish Academic Press.
Rae, M, Regan, H & Stenhouse, J (eds) (1994) *Science and Theology: Questions at the interface*, T & T Clark, Edinburgh.
Sharpe, KJ (1984) *From Science to an Adequate Mythology*, Interface Press, Auckland.

3 EVOLUTION AS HISTORY

Ayala, FJ (1998) In Russell (p 27; see citation in Chapter 5).
Hawking, SW (1988) *A Brief History of Time*, Bantam Books, London.
Johnson, PE (1991) *Darwin on Trial*, Regnery Gateway, Washington.
Livingstone, DN (1987) *Darwin's Forgotten Defenders, The encounter between evangelical theology and evolutionary thought*, WB Eerdmans, Grand Rapids, Michigan.
Plimer, I (1994) *Telling Lies for God: Reason vs creationism*, Random House Australia, Sydney.
Schroeder, GL (1990) *Genesis and the Big Bang: The discovery of harmony between modern science and the Bible*, Bantam, New York. This is an interesting book viewing the issues from a Jewish perspective.
Selkirk, DR & Burrows, FJ (1987) *Confronting Creationism: Defending Darwin*, UNSW Press, Sydney.
Stearns, SC & Hoekstra, RF (2000) *Evolution: An introduction*, Oxford University Press, Oxford.
Williams, GC (1996) *Plan and Purpose in Nature*, Weidenfeld & Nicolson, London. An excellent and clear summary of the principles of evolutionary biology.

Weinberg, S (1977) *The First Three Minutes*, André Deutsch, London. Popular introduction to the events immediately after the big bang.

4 EVOLUTION AS PROCESS

Alexander, RD (1987) *The Biology of Moral Systems*, Aldine de Gruyter, New York.
Behe, MJ (1996) *Darwin's Black Box: The biochemical challenge to evolution*, The Free Press, New York.
Birch, C (1990) *On Purpose*, UNSW Press, Sydney.
Birch, C & Cobb, JB (1981) *The Liberation of Life*, Cambridge University Press, Cambridge. A view of biological evolution from the perspectives of an eminent biologist and a process theologian.
Brack, A (ed.) (1998) *The Molecular Origins of Life: Assembling pieces of the puzzle*, Cambridge University Press, Cambridge. A convenient introduction to the 'state of play' in this confusing area.
Collinge, NE (1993) *Introduction to Primate Behaviour*, Kendall Hunt, Dubuque. Describes the Macaque cultural studies.
Davies, P (1983) *God and the New Physics*, Penguin, London.
— (1995) *The Cosmic Blueprint*, Penguin, London.
Dawkins, R (1978) *The Selfish Gene*, Granada, London.
— (1991) *The Blind Watchmaker*, Penguin, London. A fascinating materialist view of creation.

5 CREATION AND EVOLUTION

Atkins, P (1994) *Creation Revisited*, Penguin, London. One of the classic materialistic views of creation.
Berry, RJ (1996) *God & the Biologist: Faith at the frontiers of science*, Apollos, Leicester. An evangelical approach to evolution.
Gilkey, L (1965) *Maker of Heaven and Earth: study of the Christian doctrine of creation*, Anchor Books, New York. One of the classic considerations of the subject.
Holloway, R (1990) *The Divine Risk*, Darton, Longman and Todd, London.
Meyendorff, J (1979) *Byzantine Theology: Historical trends and doctrinal themes*, Second Edition, Fordham University Press, New York. A splendid and readable introduction to the world and beliefs of the early Eastern church.
Peacocke, AR (1979) *Creation and the World of Science*, Clarendon Press, Oxford. Standard work on the subject.
Peters, T (ed.) (1998) *Science and Theology: The new consonance*, Westview Press, Boulder, Colorado.
Phillips, JB (1952) *Your God Is Too Small*, Epworth Press, Peterborough.
Polkinghorne, J (1988) *Science and Creation: The search for understanding*, SPCK, London.
— (1994) *Science & Christian Belief: Theological reflections of a bottom-up thinker*, SPCK, London.

Russell, RJ, Stoeger, WR & Ayala, FJ (eds) (1998) *Evolutionary and Molecular Biology: Scientific perspectives on divine action*, Vatican Observatory, Vatican City. This is an excellent and thorough introduction to many of the scientific and theological issues raised in this book. It introduces the ideas of several different theologians.

Ward, K (1996) *God, Chance & Necessity*, One World, Oxford. A masterful presentation of natural theology by a leading theologian.

Weinberg, S (1992) *Dreams of a Final Theory*, Pantheon, New York. A materialist view of creation.

6 REDEMPTION: FROM WHAT, TO WHAT?

Many of the books listed for Chapter 5 also cover issues raised in this Chapter.

Basil of Caesarea given in Prestige, GL (1952) *God in Patristic Thought*, SPCK, London.

Daly, G (1988) *Creation and Redemption*, Gill and MacMillan, Dublin. A thoughtful and insightful book.

Thomas Aquinas: In Fox, M & Sheldrake, R (1997) *Natural Grace*, Image Books, New York, (p. 30). Not a recommended book.

Polkinghorne, J (1989) *Science and Providence: God's interaction with the world*, SPCK, London.

7 OUR PLACE IN CREATION

Bouma-Prediger, S (ed.) (1995) *The Greening of Theology*, Scholars Press, Atlanta.

Bradley, I (1990) *God is Green: Christianity and the environment*, Darton, Longman and Todd, London.

Collins, P (1995) *God's Earth: Religion as if it really mattered*, Dove, Melbourne.

Fox, M (1983) *Original Blessing*, Bear, Santa Fe. A 'new age' Christian approach to the environment.

Fox, M (1983) *Meditations with Meister Eckhart*, Bear, Santa Fe.

Hall, DJ (1990) *The Steward: A biblical symbol come of age*, WB Eerdmans Press, Grand Rapids, Michigan.

Horton, D (2000) *The Pure State of Nature: Sacred cows, destructive myths and the environment*, Allen & Unwin, Sydney. A totally different view of our relationship with nature in the Australian context.

IUCN/UNEP/WWF (1991) *World Conservation Strategy*, IUCN, Gland, Switzerland.

Lilbourne, GR (1989) *A Sense of Place: A Christian theology of the land*, Abingdon Press, Nashville.

Limouris, G (1990) *Justice, Peace and the Integrity of Creation: Insights from orthodoxy*, WCC Publications, Geneva.

Maguire, DC & Rasmussen, LL (1998) *Ethics for a Small Planet*, SUNY Press, New York.

New, TR (2000) *Conservation Biology: An introduction for southern Australia*, Oxford University Press, Melbourne.

DISCUSSION STARTERS

Reflection on issues central to one's beliefs is often aided by comparing ideas and critical assessments of ideas with others. To this end I have included a set of questions that can be used either as starting points for independent reflection or as discussion starters in a reading group.

CHAPTER 1

1 Was the description of science given here similar to those you had encountered before reading this chapter? If not, in what ways did it differ?
2 Does an image of science as the product, not of gods, but of very human individuals with very human frailties make you feel better or worse?
3 Does the fact that science describes approximations rather than the 'real' laws relieve you or disappoint you?

CHAPTER 2

1 With what confidence, and under what conditions, do you think reason can be used when considering the revelation of God?
2 What evidences available to the individual or to the community are acceptable data for theological reflection and analysis?
3 Are observations of nature acceptable data for theological reflection? Are the descriptions given by science as to the way the world works acceptable data for theological reflection?

CHAPTER 3

1. The history of the universe places humankind on a small planet circling a minor sun at a not distinguishable moment in the 12 to 15 billion-year history of the universe. Does this make any difference to your value or worth to yourself or God?
2. Does the fact of your origins, as made from 'star stuff', with a long line of animal ancestors, change your perception of yourself, of your perception of your relationship to the natural world; of your perception of your relationship to God?
3. Modern physics, in a similar fashion to Augustine, sees God as being 'out' of time. The moment of the big bang, Christ's death and the end of the present universe are all present in God's knowledge as a single unitary item. There are no afterthoughts in God's actions and forgiveness for past sins is given in the full knowledge of future failings. How does this perspective on human history affect your understanding of Christ's action?

CHAPTER 4

1. If you are the results of a process, or only exist as a process then what are you?
2. If you are the result of your context, relationships and history, who are you?
3. If we think for a moment about brain-dead individuals (though physiologically functional), we are faced with the question of what constitutes 'alive' and 'dead' in the human sense in the most confronting form. What does 'alive' mean to you for a plant, a wild animal and a human?

CHAPTER 5

1. If exploration not progress is the purpose of evolution, how does this affect the way we should live?
2. If meaning is to be found in 'being what is intended' and, if the maintenance and deepening of relationships is implicit in the structure of the universe, how should this affect our values and priorities?
3. What could we replace 'material progress' with as a basis for our society?

CHAPTER 6

1. For the Eastern Fathers, meaning and purpose in life are found by being truly ourselves and we can only be truly ourselves if the pattern of our relationship with God is as Christ, our initiator and our fulfilment, intended. Does such an image match your experience of meaning in your faith?
2. To understand life as a process, rather than achieving a set of goals or passing certain exams set for us by life, is to leave us without measures of success or failure. What, then, are we to make of the term, 'saved'?

3   Does the concept that Christ has destroyed death in some ultimate sense, make a difference to you, given that you are still going to grow old and die in the world's terms?

CHAPTER 7

1   How does the concept that the cosmos is more than just a backdrop to your life affect you?
2   If meaning is to be found in being truly ourselves, then how would accepting your place in the world as priest and steward affect the meaning and value you currently place on your life?
3   In what ways would you need to change your behaviour if you accepted these roles? How would you differentiate between establishing a high quality of life and a high standard of living?

INDEX

adaptation 51, 53
aging 59
alienation 80, 82
altruism
  evolution of 55
anthropic principle 43, 52
apathy 76
Ardipithecus 38
Assumptions
  auxiliary 6
Australopithecus 38

baptism 91
Basil of Caesarea 85
beauty 65
big bang 27, 29
biological diversity 95

categories 84
chance 83, 111
chaos 66, 67
choice 57, 111

Christ 72, 87, 89, 91, 92, 104,
competition 50, 60, 79, 90
complexity
  biochemical 53
constants
  fundamental 43
creation 72-77, 109-114
Creator 61, 67, 72, 73, 80, 97, 102, 106
crisis
  environmental 94
cyanobacteria 36

Darwin, C, 51
death 49, 57, 84, 86, 87, 88
deduction 7
dendrochronology 27
design
  argument from 51, 52, 53, 70
  imperfections in 54
DNA 33, 47
  human 39
dominant paradigm 12

dominion
  over nature 106
duplication
  gene 48

Eastern Church Fathers 84
economic rationalism 75
El Niño 67
elegance 3, 43, 65
emergent properties 3, 79, 89, 90, 97, 98
enabling 70, 71, 72, 99, 105, 107
entropy 28, 32, 58
evil 58, 83
evolution 35, 71, 97
  as history 23-40
  as process 51-62
  assumptions of 22
  of cultures 57
eye 52-54

fall 84, 86
falsification 8, 11
Fathers
  Eastern Church 84
fear 78
fitness 49
  inclusive 54
flood
  worldwide 25
fossils
  of intermediate form 37
  oldest 36
freedom 65, 66-68, 70-73, 80-83, 85, 88, 99, 113
futility 69, 71, 78, 83

Genesis 110
gravity 30

Hildegard of Bingen 96
history 36
history
  human 38
Holy Spirit 72, 104, 106
Homo erectus 39
Homo ergaster 39

Homo sapiens 39
humanity
  history of 38
Huxley, TH, x

imagination 6
incarnation 88
induction 5
innovation 49
intelligibility 71
intergenerational equity 100
intuition 5

Jesus 18, 73, 74, 87, 88, 90, 93, 98
Job, Book of 83
justification *see* verification

kin selection 55

laws
  of nature 43, 85
life
  evolution of 35
  origins of 32, 46
liturgy
  patristic 102
logic 7
love 70, 71
Lucy 38

materialism 75
  and science 64
mathematical descriptions 3
Maximus the Confessor 88
meaning 69, 70, 74
Meister Eckhart 96, 103
micro-evolution 51
Monod, J, 83
morality 98
mutagens 47
mutation 52
mutations 47, 48, 49
myth 100, 109

natural theology 19-21, 78
nature 95-108
  behaviour towards 101

dominion over 106
  enabling 99
necessity 83
neighbour 98, 101
Noah 25
Noahian flood 25

Occam's razor 9, 64
old age 59
origin
  of species 51
  of life 46
overpopulation 58, 95

paradigm shift 4
parsimony 9
patristic period 84
patristic view 85, 88
physics
  quantum 3, 42, 45, 48, 66
postmodernism viii, 9, 21, 75
preconceptions 6, 7, 30
pride 76
priestly role 102, 107
progress 60, 69, 96, 110
  myth of 75
properties
  emergent 3, 79, 89, 90, 97, 98
proteins 33
pseudoscience 11
purpose 20, 65, 66-74, 76, 97

rationality 65
reason 17, 20
redemption 78-93
relationships 90
  in ecology 46
  in humans 46
  in quantum physics 45
religion
  philosophy of 16
repentance
  and nature 101
reverence 105
RNA 33

salvation 78, 88

'saved' 81
science 1-14, 18, 20, 30, 41, 45, 63, 64, 69, 70, 76, 78, 107
  and Christianity 85
  as analogy 11
  assumptions of 1, 2
  definition of 1
  limits of 14
  process of 5
scientism 10, 64, 75
scientists
  as a community 11
  image of 1, 9
selection
  hard 49
  kin 55
  natural 35, 47, 48, 49, 52, 54
  soft 50
selfishness 56
senescence 59
simplicity
  exclusive 64
  inclusive 65
sin 80, 86, 92
speciation 51
spirituality
  Aboriginal 103
St Augustine 84
stewardship 106
stromatolites 36
suffering x, 57, 58, 83

theology 15-22
  assumptions of 16
  definition of 16
  limits of 21
  natural 19, 63
theory 8, 12, 14
  satisfactory 13
thermodynamics
  second law of 28, 32, 49
time 27, 30
truth 71

uncertainty
  quantum 11, 67
uniformitarianism 24

value 65
  aesthetic 96
  intrinsic 97
  utilitarian 95, 100
verification 5, 7, 19
  intersubjective 2

Ward, K, 64, 65

watchmaker 52, 53, 70
Western Church 85
Wisdom 106, 111ff
World Conservation Strategy
  100
worship 102, 103

young earth explanations 25